COVER UP

COVER UP

WHAT THE GOVERNMENT IS STILL HIDING ABOUT THE WAR ON TERROR

PETER LANCE

HARPER

NEW YORK . LONDON . TORONTO . SYDNEY

HARPER

P. 263: "The Jersey Girls": Suchat Pederson; Beverly Eckert: Peter Lance; Monica Gabrielle: Peter Lance; 264: Mohammed Ajaj: AP Photo/Rick Bowmer; 265: Khalid Shaikh Mohammed: AP Wide World; 266: Larry Mazza: Charles A. Arrigo/New York Daily News; Joe Simone: Peter Lance; Angela Clemente: Peter Lance; 267: Louis Freeh: AP Photo/Dennis Cook; Aida Fariscal: Peter Lance; Rodolfo Mendoza: Peter Lance; Valerie Caproni: AFP/Lucy Nicholson/Getty Images; James Kallstrom: AP Photo/Mark Lennihan; 268: Gov. Thomas Kean, Rep. Lee Hamilton, Sen. Slade Gorton, Richard Ben-Veniste, Sen. Bob Kerrey, John Lehman, Jamie Gorelick, Fred Fielding, Rep. Tim Roemer, Gov. James Thompson: 9-11commission.com; Philip Zelikow: MCPA; Sen. Max Cleland: AP Photo/Dave Martin; 269: George W. Bush, 9/11/01: Win McNamee/Reuters; Condoleezza Rice: AP Photo/J. Scott Applewhite; Richard Clarke: AP Photo/Dennis Cook; 271: WTC, 9:02 A.M., 9/11/01: AP Photo/Carmen Taylor; 272: The Pentagon, 9/11/01: AP Photo/*The Daily Progress,* Dan Lopez; Shanksville, Pennsylvania, 9/11/01: AP Photo/*The Plain Dealer,* Joshua Gunter.

A hardcover edition of this book was published in 2004 by ReganBooks, an imprint of HarperCollins Publishers.

FIRST HARPER PAPERBACK PUBLISHED 2005.

Designer: Publications Development Company of Texas

The Library of Congress has catalogued the hardcover edition as follows:

Lance, Peter.
Cover up : what the government is still hiding about the war on terror / Peter Lance.—1st ed.
p. cm.
Includes bibliographical references and index.
ISBN 978-0-06-054355-6
1. Terrorism—Government policy—United States. 2. Intelligence service—United States. 3. Official secrets—United States. 4. National Commission on Terrorist Attacks upon the United States. 5. Septembre 11 Terrorist Attacks, 2001. I. Title.

HV6431.L343 2004
973.931—dc22

2004051326

ISBN 978-0-06-079511-5 (pbk.)

09 10 11 12 13 PDC/QWF 10 9 8 7 6 5 4 3 2

To Joseph James Lance, my father,
whose service to our country inspired me and whose
tireless work ethic forged the tenacity that
kept me on track as I sought to tell this story.

CONTENTS

INTRODUCTION

Within hours after the attacks of September 11, 2001, I began an investigation into the intelligence failures leading up to the day that would become known forever as 9/11. Almost two years later, the first phase of my work culminated in the publication of *1000 Years for Revenge: International Terrorism and the FBI—The Untold Story.* The central finding of the book was that Ramzi Ahmed Yousef, the original World Trade Center bomber, had set the 9/11 plot in motion in the Philippines as early as 1994. Along with his uncle Khalid Shaikh Mohammed, Yousef had targeted six buildings on both U.S. coasts, including the Twin Towers, the Pentagon, CIA headquarters, the Sears and Transamerica Towers, and a nuclear facility. Furthermore, as demonstrated by evidence I obtained from the Philippines National Police official who unearthed the plot, by 1995 there were as many as ten Islamic radicals training in U.S. flight schools.

Finally, my investigation proved that the PNP had given this evidence to the U.S. Embassy in Manila—and that it had been received by the FBI. And yet, after Yousef and two of his cohorts were captured, a Bureau agent allowed Khalid Shaikh Mohammed to escape. A year later, while Yousef and two other al Qaeda operatives were being tried in New York for the Manila air bombing plot they called "Bojinka," Khalid Shaikh Mohammed went on to perfect the hijacked-airliners suicide plot that his nephew Yousef had set in motion in 1994—the plot that became 9/11. But as I discovered in my investigation, the FBI and the Justice Department had failed to follow up on the Philippines evidence, and the hunt for Khalid Shaikh Mohammed himself was shrouded in secrecy.[1] The question was, why?

Within weeks of publication, *1000 Years for Revenge* was used by Max Cleland, one of the original 9/11 commissioners,[2] to kick off questioning in the afternoon session of the Commission hearings in Washington, D.C., on October 14, 2003.[3]

Six months later, I presented a summary of my findings to the 9/11 Commission itself in testimony before one of its senior attorneys and a staff investigator in New York. In a prepared statement, which I followed up with documentary evidence, I summarized the key findings presented in the book, and raised a series of unanswered questions that I hoped the Commission would address in its own probe into the greatest mass murder in U.S. history.* In its enabling legislation, the goal set for the 9/11 Commission by Congress was to pick up where the Congressional Joint Inquiry into the 9/11 attacks had left off—and "build upon" its findings.[4] But the Joint Senate-House investigation was severely criticized by its co-chairman, Sen. Richard Shelby (R-AL), for failing to assess accountability.[5] It was the hope of the organized 9/11 victims' rights community that this new panel, formally known as the National Commission on Terrorists Attacks upon the United States, would advance the Joint Inquiry's work and get to real truth.

Now, almost two years after its creation, the core group of the 9/11 Family Steering Committee—Patty Casazza, Lorie Van Auken, Mindy Kleinberg and Kristen Breitweiser—believes that the 9/11 Commission has failed in its mission when it comes to assessing blame.

Known in the media as the "Jersey Girls," these four widows (along with Monica Gabrielle of Manhattan and Beverly Eckert of Stamford, Connecticut),† more than any other single group, used their collective tenacity and heart to lobby the 9/11 Commission into being in the first place.[6] In the past twenty-two months they have fought for the Commission's funding, and for an extension of time for its investigation. They have demanded integrity from the Commission's staff, and insisted on the full disclosure of documents that would help shed some light on why their loved ones, along with 2,968 others, had to perish on that day.[7]

* My testimony before the 9/11 Commission is contained in Appendix I on pp. 273–293.

† See picture, p. 263.

The Commission's final report came out at the end of July 2004. And yet, as early as late May, Lori van Auken—whose husband Ken died in the World Trade Center's North Tower—told me that the Commission had breached its responsibility. "We feel that the Commission already has its report written," she said. "It's our sense today that they decided early on what they wanted the public to know and then geared the hearings to fit this pre-conceived script. That's why they raised very few of the key questions that we asked them to raise."*[8]

1000 Years for Revenge began with a quote from the Roman poet Juvenal: "Who is guarding the guardians themselves?" It is clear now, after a nine-month examination of the Commission's work, that its final report fails to address some of the most fundamental questions about the intelligence failures that led up to the attack, and the catastrophic defense failures that occurred on September 11 itself. The 9/11 Commission, in short, is destined to go down as the Warren Commission of its era—an official body that purposely limited the scope of its investigation, cherry-picked evidence, and allowed political considerations on the left and the right to influence its final conclusions.

More important, as evidenced by Commissioner Jamie Gorelick's admission in April 2004 that "the vast preponderance of our work, including with regard to the Department of Justice, focuses on the period of 1998 forward,"[9] it is clear that the Commission ignored vast areas of culpability in the U.S. intelligence community. In *1000 Years for Revenge,* I noted that the twelve years leading up to 9/11 represented the greatest intelligence failure since the Trojan Horse.[10] For that negligence, three separate presidential administrations were to blame. But the catastrophic mistakes that took place on 9/11 itself also represented the greatest *defense* failure in American history.

At the end of this book, I'll offer a detailed, minute-by-minute analysis of the seeming inability of the White House, NORAD, and the FAA to interdict the attacks at their inception, and minimize the potential damage the terrorists would cause. For authority I'll reference excerpts from the

*Those questions can be accessed at the web site of the Family Steering Committee, http://www.911independentcommission.org.

extraordinary research of Paul Thompson and the Center for Cooperative Research. Available at www.cooperativeresearch.org, Thompson's work includes a series of timelines, dating back as early as 1979, that track the emerging Islamic terror threat in precise detail. Ironically, with almost no budget and simple access to the Internet, Thompson has done what the 9/11 Commission failed to do, fulfilling the intent of Congress to "make a full and complete accounting of the circumstances surrounding the attacks."[11] Further, he's supported his findings with mainstream media sources in what amounts to an almost week-by-week history of the events that preceded the attacks.

When the dots on that terrible day are lined up, the reader will be confronted with a sobering illustration of government negligence at its worst. An analysis of the remaining unanswered questions from 9/11 will serve to underscore the inadequacy of the Commission's report, which the victims community had hoped would provide both an honest account of intelligence shortcomings and a blueprint for change.

Cover Up is intended as an unofficial "minority report" on the shortcomings of the Commission's work. It is my hope that after examining this book, the Cooperative Research timelines, and the work of the Family Steering Committee, lawmakers will be persuaded to insist on the kind of unbiased and complete investigation that must be performed if this country is ever to be protected effectively from the growing threat of international terrorism.

Patty Casazza, who lost her husband John in the South Tower, expressed the fears of many Americans after the New York hearings in May with a few simple words: "A lot of questions that the public has, may die with this Commission."[12]

The intent of this book is to keep those questions alive.

Al Qaeda and the Crash of TWA 800

I began this phase of the investigation in the fall of 2003, shortly after publication of *1000 Years for Revenge.* Like any investigative reporter, I started kicking over rocks, and developing sources inside the Commission.

Within months, the probe had unearthed startling new evidence that helped me learn the truth behind the central unanswered questions of my last book:

- Why was the evidence that Ramzi Yousef had hatched the 9/11 plot in the Philippines back in 1994 seemingly ignored by the FBI?

- Why did the Justice Department narrow the scope of the Bojinka trial of Yousef and two other al Qaeda cohorts in 1996—so that the name of the bomb maker's uncle, Khalid Shaikh Mohammed, "the mastermind of 9/11," was never mentioned in more than three months of testimony?

- Why did the FBI and the Justice Department keep the hunt for Khalid Shaikh Mohammed—a top al Qaeda operative with direct access to Osama bin Laden—secret for years, and fail to utilize the same public hue and cry that had brought his deadly nephew Ramzi to ground?

In effect, the Bureau and the Justice Department were alerted to the 9/11 plot—and the identity of Khalid Shaikh Mohammed, its chief executioner—more than six years before September 11, 2001. Why didn't they stop him? This time I found not only the answers to those questions, but shocking proof of al Qaeda's ties to the *second* biggest mass murder in American history: the downing of TWA Flight 800 in 1996, which claimed two hundred and thirty lives. Evidence uncovered for this book now shows persuasively that the FBI was warned, weeks in advance of the crash, that al Qaeda operatives tied to Ramzi Yousef would explode a bomb aboard a U.S. airliner in order to bring about a mistrial in the ongoing Bojinka case.

At the time, Federal agents were actually given an opportunity to meet with active al Qaeda cell members here and abroad—an initiative that might have led to the capture of 9/11 mastermind Khalid Shaikh Mohammed—but rejected the offer. Worse yet, the evidence suggests that the failure of an FBI intelligence-gathering operation, set up to monitor Yousef, may have even facilitated the TWA 800 crash.

In order to grasp these early FBI missteps as foreshadowings of 9/11, it's important first to understand how the Bureau handled the intelligence it received in the months leading up to the TWA crash on the night of July 17, 1996. Toward that end, in Part One of *Cover Up* we'll examine in depth the FBI's original criminal investigation of the crash, which was aborted for reasons explained here for the first time. We'll present new evidence that calls into question the Bureau's official explanation for the presence of high explosives aboard the wreckage of the Paris-bound flight. And we'll focus on a series of unrelated criminal cases, which we believe led the FBI and the Justice Department to hide the actual cause of TWA 800's crash.

The cover-up itself was the result of an ends/means decision made at the upper echelons of the Clinton Justice Department. In the process, a major event in the war on terror was ignored in order to keep a series of organized crime convictions from coming unraveled. But two of the central figures involved hold top law enforcement posts in the Bush Administration today. Perhaps most troubling of all, the staff of the 9/11 Commission was given notice of this evidence in April 2004—and chose to dismiss it.[13]

In its mission to solve the greatest mass murder in U.S. history, the 9/11 Commission has ignored evidence indicating that as far back as 1996, the FBI turned down a chance to monitor an active al Qaeda cell in New York City, solve the TWA 800 crash, and connect the dots all the way to Khalid Shaikh Mohammed as he plotted the 9/11 attacks in hotel rooms, flight schools, and tenement apartments from Doha, Qatar, to Hamburg, Germany.

Because much of the evidence presented in *Cover Up* is visual, we'll present it in a series of illustrated appendices at the end of this book. The rest of the supporting data can be found at www.peterlance.com—including dozens of FBI 302 memos, and the extraordinary reports of an FBI confidential informant, never before seen outside the FBI and the Justice Department.

Although the informant reports were later disavowed by U.S. attorneys and a federal judge, the material offers an extraordinary glimpse inside the mind of Ramzi Yousef, the most dangerous single terrorist the United States has ever faced. They not only predict the TWA 800 crash, but offer a dire warning of the al Qaeda hijackings and violence that would face America and the rest of the world years later. One of the most startling revelations in the 9/11 Commission's final report was the disclosure of a 1998 Presidential Daily Briefing warning that al Qaeda-related terrorists would hijack a plane to free the so-called "blind Sheikh," Omar Abdel Rahman. In *Cover Up,* for the first time, we'll present an FBI 302 memo showing that the Bureau and the Justice Department were warned of an identical threat *as early as 1996.* If acted upon with dispatch back then, that informant report might have led federal investigators to interdict the 9/11 plot years before 2001.

This is the story of an extraordinary act of negligence in one presidential administration that was compounded in the next, then ignored by the 9/11 Commission, the last official investigative panel charged with finding the truth about the attacks.

In *1000 Years for Revenge,* I followed the road to 9/11 back twelve years, to the summer of 1989. The road to the solution of the TWA 800 crash, improbably enough, extends back even further. It begins in 1964, forty years ago this past summer, when Ku Klux Klan killers hid the bodies of three civil rights workers beneath the ochre clay of an earthen dam in Meridian, Mississippi . . . and a desperate J. Edgar Hoover recruited a sociopathic Mafia killer to head south from New York City and break the case.

PART I

I don't think anybody could have predicted that these same people would take an airplane and slam it into the World Trade Center, take another one and slam it into the Pentagon; that they would try to use an airplane as a missile.

—National Security Advisor Condoleezza Rice, May 17, 2002

If I had any inkling whatsoever that people were going to fly airplanes into buildings, we would have moved heaven and earth to save the country.

—President George W. Bush, April 13, 2004

The Joint Inquiry confirmed that, before September 11, the Intelligence Community produced at least twelve reports over a seven-year period suggesting that terrorists might use airplanes as weapons.

—Report of the Joint Inquiry into the Terrorist Attacks of September 11, 2001
by the House Permanent Select Committee on Intelligence
and the Senate Select Committee on Intelligence

1

THE FBI'S KILLING MACHINE

The White Knights of the Ku Klux Klan who ruled Neshoba County, Mississippi, had a nickname for Michael "Mickey" Schwerner, the New York white boy who came down to Meridian that Freedom Summer to register black voters. "Goatee," they called him. Andrew Goodman, another young white man, and James Chaney, a black civil rights worker, were referred to more prosaically, as "outside agitators."[1]

On June 21, 1964, the three young men, members of the Congress of Racial Equality (CORE), had come to nearby Philadelphia, Mississippi, to investigate the Klan burning of Mt. Zion, an African American church. According to evidence presented at a federal trial years later, Edgar Ray Killen, a local preacher and KKK Kleagle, conspired with at least twenty other Klansmen to lie in wait for the three men. Arrested by Cecil Price, the town deputy, for speeding, Goodman, Schwerner, and Chaney had been jailed for five hours. Presiding over a Klan meeting that night at the Long Horn Drive-in, Killen purportedly described what should happen to the young men as "elimination."[2] But one of the Klansmen would later testify that Killen's death warrant was less delicate. "He said those . . . civil rights workers were locked up and they needed their rear ends tore up."[3]

Released after midnight, so that two cars full of Klansman could follow their progress undetected, Goodman, Schwerner, and Chaney left the Neshoba County Sheriff's office heading north from Philadelphia in a Ford station wagon. They were never seen alive again.*

Hiding behind an old warehouse, the two-car Klan convoy, followed by Deputy Price in his patrol car, hung back until Goodman, Schwerner, and Chaney cleared the outskirts of town. At that point they gave chase, with one vehicle, a red Chevy, roaring up beside the Ford wagon and attempting to force it off the road. After a pursuit on Highway 19 that topped speeds of one hundred miles an hour, Deputy Price flashed his red light, pulled the Ford over, and ordered the civil rights workers to get out. The three young men complained, holding their hands up when flashlights were pointed in their eyes. They were then driven thirty-four miles away. When the convoy turned off onto a graded clay road, Goodman and the others were ordered out of their cars again, and shot to death where they stood. Their bodies were thrown into the back of their own station wagon and driven to the farm of a wealthy Philadelphia businessman. There they were dumped at the base of an earthen dam while a bulldozer entombed them beneath fifteen feet of red clay.

Their disappearance touched off a firestorm at the Justice Department. Dozens of dark-suited FBI agents were dispatched to Neshoba County in what the Bureau dubbed the MISSBURN case.

As recreated in the 1988 film *Mississippi Burning,* the agents located the Ford, which had been taken to a local swamp area and burned. But soon the investigation stopped dead. Terrified blacks and protective whites hid that fact that the murders had been the work of twenty-one local men from the Klan Klavern.[4]

In 1964, the FBI may have been hated in Mississippi, but the Klan was *feared.* At that point the self-described "White Knights" were waging a war of domestic terror throughout the state, perpetrating church bombings, arsons and the beating of black citizens. "Nobody, and I mean *nobody,* was gonna give up those boys," recalls W. O. Chet Dillard, a Neshoba County judge who had served as district attorney several years

* See picture, p. 266.

later.[5] "Old J. Edgar figured that if he was gonna break that thing—and he was hurtin' to break it—he was gonna have to go to some extreme measures . . . and he did."

J. Edgar Hoover, the notorious FBI director who for years had denied even the existence of organized crime in America, decided to call on a man who had already earned a position in an elite group inside the Bureau known as TE for "Top Echelon."

Since 1962, Gregory Scarpa Sr.* had been living two lives.[6] A young Bensonhurst, Brooklyn, cock of the walk, Scarpa was a soldier in the Columbo crime family. But he was also working as a mole, an informant for the FBI.

Years later, Scarpa's boss, Joe Colombo, would tear up the Mafia rule book, which demanded anonymity, and found the Italian American Civil Rights League. It was a very public "Italo-American" advocacy group that regularly picketed FBI offices and staged rallies to denounce the Bureau for "Forever Bothering Italians."

In time, Greg Scarpa would become one of Colombo's flashiest spokesmen. He was also a lucrative earner, bringing in tens of thousands of dollars a month from bookmaking, loan-sharking, credit card fraud, and other racketeering operations. But he had a violent side, bragging that he loved the smell of gunpowder, and punching a Satanic "666" into the beepers of fellow wiseguys each time he made a kill.[7] And yet, on the flip side of his double life, he would sit down at least once a week with Anthony Villano, his Bureau control agent, spilling a host of secrets on the inner workers of what Hoover later dubbed the LCN—La Cosa Nostra.[8] Soon Scarpa married and had a son, whom he named after himself. But he didn't draw Greg Jr. into the family business until he was a teenager.

Dubbed "the Grim Reaper" and "the Mad Hatter" by his colleagues, Greg Scarpa Sr. became known as a *capo* who could take a fellow wiseguy to lunch, joke with him over a plate of ziti, and shoot him between the eyes before the check was paid.[9] He had one of his victims hit while he was stringing Christmas lights in front of his wife, and ordered a grave dug for another victim in advance of the whack.

*See picture, p. 266.

Later, when Joe Columbo was killed in a Columbus Circle rally in 1971, Scarpa became the principal shooter in a war of succession between rival factions of the Columbo family. At first the family was taken over by Carmine "the Snake" Persico. But when Persico went to prison, leadership passed to his son Alphonse, aka "Allie Boy," who was widely perceived as weak.[10] The war that followed was bloody; the result was a series of drive-by shootings and gun battles that played out across Brooklyn and Staten Island between 1991 and 1992. In the end, twelve were dead, including two innocent bystanders, and a third of the victims were murdered by Greg Scarpa Sr. himself.[11]

Interrogation by Razor Blade

Back in 1964, however, when the call came from Hoover, Greg Scarpa Sr. was still just a ruthless young *soldati* with ice water in his veins. The latter-day historical perception of the case has been shaped by the film *Mississippi Burning,* in which the kidnapping and triple murder is broken by an African American FBI agent who is flown down to question the town's mayor. But through previously unpublished documents from declassified FBI files,[12] interviews with Neshoba county residents, and other research, for the first time the real story has finally come to light: In truth, it was Greg Scarpa Sr., using his talents as an enforcer, who broke the Goodman/Schwerner/Chaney case.

Having been authorized by Hoover for what the FBI called the "special," Greg flew with his seventeen-year-old girlfriend, Linda Diana, to Mobile, Alabama, where he was met by FBI agents who gave him a gun. According to biographer Sandra Harmon, who spent months interviewing Linda for a book, Scarpa told her: "I'm going out with these men. If I'm not back, there's a return ticket and here's some money," passing her a folded stack of bills.[13]

Declassified FBI documents and the files of Judge Dillard, who was then Neshoba County DA, show that Scarpa was then driven to Philadelphia, Mississippi, where, with the help of FBI agents, he kidnapped a local politician.

"They took him to an undisclosed location," said Judge Dillard, "and while the agents waited outside, Scarpa started working on the guy. He put a pistol to his head and cocked it, demanding to know where those boys were. But the man told him a phony story. After he checked it with the agents, Scarpa then put the barrel of the gun in the man's mouth and cocked it."

Fearing Klan reprisal, though, the audacious politician gave Scarpa another story; again, the agents outside confirmed was false. "It was at that point," says Judge Dillard, "that Scarpa took more drastic steps."

Taking out a straight razor, he proceeded to unzip the man's fly. "He was threatenin' to emasculate him," says the Judge. "And that's when he blurted out the location of the dam."

It wasn't the only time the Feds would call upon Scarpa for this kind of assignment. Eighteen months later Hoover recruited the mafioso again, this time to break a case involving the Klan's firebombing of NAACP worker Vernon Dahmer's home, in which Dahmer was killed and his wife and daughter were badly burned. Though local witnesses stood mute, the FBI's investigation led to Lawrence Byrd, a Klan captain who owned an appliance store in nearby Laurel, Mississippi.

As Judge Dillard recalled, Scarpa and two other men wearing wigs showed up at the appliance store just before closing one night in January 1966. "They showed an interest in some television units, but as [Byrd] came around the counter to show them, they grabbed him, hijacked him into an automobile [and] took him out on a back road."[14]

There, according to Judge Dillard—who later interviewed the terrified Byrd in the Jones County Hospital—Scarpa proceeded to "beat him within an inch of his life."

"They threatened to string him up and leave him out there naked in the winter, where the animals could get at him," said the judge. "After that I can tell you that Mr. Byrd told them what they needed to know, and later on it's pretty clear that he set up another man related to the Dahmer case who was also kidnapped and beaten into a confession."[15]

"Lawrence was a tough guy," Dillard told a writer for the *New Yorker* in 1996, "a big rawboned country boy, but he was beat up so bad, that he was never the same after that."[16]

On returning from that second "mission," as Scarpa called it, he and Linda vacationed at the Fontainebleau Hotel in Miami, courtesy of Uncle Sam. Several accounts of Scarpa's work for the FBI have mixed the two missions together.[17] But Judge Dillard insists that Byrd's confession was related to the Dahmer case only.

In any event, though seven Klansmen were convicted of federal conspiracy charges in connection with the Goodman/Schwerner/Chaney slayings, the reputed leader of the ring, Rev. Killen, walked out of court a free man and none of the conspirators was ever even indicted on murder charges.

On its fortieth anniversary, the case remains a kind of legal open wound in Mississippi. In late May, due in part to the reporting of the *Neshoba Democrat* and the *Clarion Ledger,* Mississippi Attorney General Jim Hood announced that he was reviewing more than forty thousand pages of FBI documents, with an eye toward possibly reopening the case.[18]

"All these years I have been hoping there would be justice for those boys," says Caroline Goodman, whose son Andrew was twenty at the time of his murder. "I'm still waiting."[19] In that case, the U.S. Justice Department, which is supporting the new investigation, is on the side of the angels.

Decades after Greg Scarpa's Mississippi black-bag job, however, the FBI's involvement in the darker side of his Bureau service led to an extraordinary obstruction of justice in a very different setting—the war on terror.

Getting Married to the Girlfriend

Many years before Gregory Scarpa Sr.'s death in 1994 Special Agent Anthony Villano had been forced to "close" him as a confidential informant (CI). FBI rules forbid the use of CIs if there is even a hint that they're involved in a "serious act of violence." By 1970, Scarpa was suspected in the murder of a DEA informant, and while Villano "did a tap dance" to protect him and Greg escaped indictment, he was eventually shut down in 1975.[20]

Five years later, however, R. Lindley DeVecchio,* an FBI special agent working in the Bureau's New York office, noticed Scarpa's name in the closed files and re-activated him. Villano compared their fourteen-year relationship to a couple "getting married," and their partnership eventually helped both figures advance in their careers. As Scarpa took de facto control of one of the two rival branches of the Columbo family and DeVecchio was promoted to Supervisory Special Agent, heading the Columbo Task Force, he became one of the Bureau's most celebrated authorities on organized crime.

"The problem was," says defense attorney Alan Futerfas, "the relationship was deeply corrupt."[21] In fact, as the FBI's own files would later reveal, Scarpa wasn't merely defending the Persico faction in the Colombo war—he had instigated the violence *himself.* Killing wiseguys on both sides, Scarpa was content to keep the blood flowing long enough to gain control of the family himself, and grab a seat on the national "Commission" of La Cosa Nostra.

As evidence from a series of federal trials now suggests, Special Agent DeVecchio, whom Scarpa referred to variously as "Mr. Dello," "Del," and "the girlfriend," actually helped Scarpa locate his victims. According to Scarpa's oldest son, Gregory Scarpa Jr., DeVecchio even sat watch as a lookout during a bank robbery.[22] Scarpa's number two, Larry Mazza, later told investigators that DeVecchio had ordered FBI agents and NYPD cops away from a surveillance site so that Scarpa senior could blow the head off one of his chief rivals.[23]

According to the testimony of Gregory Scarpa Jr., who eventually kept the books for his father, DeVecchio was paid more than $100,000 in return for this inside help over the years. He was rewarded with vacations to places like Aspen, Colorado, and treated to call girls in a Staten Island hotel.[24]

DeVecchio, who vehemently denied the corruption charges, became the subject of a two-year FBI OPR (internal affairs investigation). Yet he was ultimately cleared of the charges, and allowed to retire with a full pension. Still he refused to take an FBI polygraph, and in a May 1996 habeas corpus hearing for a group of Scarpa's rival mobsters, DeVecchio invoked the Fifth Amendment repeatedly. Even after being granted immunity by

*See picture, p. 266.

the Government, the former supervisory special agent answered "I don't recall" forty-four times.

Further, FBI 302 memos obtained in this investigation show that three of the agents working directly under SSA DeVecchio testified internally that he had crossed the line. The allegations of DeVecchio's immediate subordinate, Special Agent Christopher Favo,[25] and agents Howard Leadbetter and Jeffrey Tomlinson dovetail with what a series of Columbo family members told investigators when they agreed to testify: that DeVecchio had not only leaked intelligence to Scarpa, he had allowed him to view FBI surveillance tapes, warned him of impending arrests, and furnished him with Bureau forensic evidence so that he could locate his opponents and hunt them down.[26]

DeVecchio himself had escaped trial on gun-running charges in 1997 when a federal judge decided not to prosecute him for illegally selling $60,000 worth of weapons in Maryland, and lying to agents of the Bureau of Alcohol, Tobacco and Firearms.[27]

Even worse, DeVecchio had intervened at least twice after Scarpa was arrested for a massive bogus credit-card scam and later for gun possession, making sure that his prized informant got probation and stayed on the street. All of this, again, was in defiance of FBI informant rules. But DeVecchio later testified that Scarpa's violent murder spree was known to the New York Special Agent in Charge, as well as top FBI officials in Washington.[28]

Between 1962, when he first began working as an FBI informant, until he stopped working for the Bureau thirty years later, Gregory Scarpa was paid upwards of $158,000 by the Feds. But that was insignificant when compared to the millions he earned from loan sharking, drug-dealing, credit-card and securities fraud, and murder for hire. In the three decades he worked under the Bureau's protective umbrella, Scarpa, also known as "the Killing Machine," did only thirty days in jail before he was ultimately sentenced in the Colombo war homicides and died in prison.

When that tangled web became unraveled in the mid-1990s and DeVecchio was exposed, it threatened to derail at least nine major homicide and RICO cases that had made the careers of a series of prominent Feds.[29]

When they were faced with that level of embarrassment, the evidence now shows that officials of the FBI and Justice Department cut their losses by shutting down the most important investigation into an act of al Qaeda violence on U.S. soil prior to 9/11. It was an intelligence lead, that if pursued, would have led the Feds right into the heart of the 9/11 plot.

2

THE MOZART OF TERROR

Until the capture of his uncle Khalid Shaikh Mohammed,[1] the man *Newsweek* dubbed "America's most wanted" terrorist[2] was Abdul Basit Mahmud Abdul Karim, aka Azan Muhammed, Adam Ali Qasim, Naji Haddad, Dr. Paul Vijay, Dr. Adel Sabah, Arnaldo Forlani, Muhammad Ali Baloch, Adam Baloch, Kamal Ibraham, Abraham Kamal, Khurram Khan, Adam Ali, Dr. Alex Hume, and Dr. Richard Smith.*[3] His family came from the no-man's-land of Baluchistan, which intersects Afghanistan, Pakistan, and Iran. After coming of age in the Kuwaiti oil city of Fuhayhil, he got his engineering degree in the United Kingdom, and soon left for the incipient al Qaeda training camps of Peshawar on the Afghan border. It was there that he first learned how to build improvised explosives devices (IEDs), and adopted the nom de guerre by which he would become known around the world: Ramzi Yousef.[4]

A master of at least six languages including English, Arabic, Urdu, and Baluchi,[5] Yousef was soon drafted as Osama bin Laden's chief point man on terror. What followed was a three-year terror spree that included not only the World Trade Center bombing in 1993[6] but plots to kill the Pakistani prime minister,[7] the Pope, and President Clinton; the creation of explosive devices designed to kill thousands of Muslims, Christians, and Jews;[8] the notorious Bojinka scenario, meant to detonate aboard eleven U.S.-bound jumbo jets;[9] and his penultimate homicidal plan: the suicide-hijacking plot that culminated on 9/11.[10]

Characterized as "cold blooded," "diabolical," and an "evil genius"[11] by those who tracked his career, Yousef was an outside-the-box thinker who went from designing huge weapons of mass destruction intended to be moved to their targets by truck,[12] to small, ingenious "bomb triggers," intended to be planted aboard commercial airliners, using the fuel tanks of the jumbo jets themselves to create enormous "flying bombs."[13]

1000 Years for Revenge presented evidence proving that Ramzi Yousef was the al Qaeda link between both attacks on the Twin Towers, and that in both acts of terror he was part of a cell funded by Osama bin Laden. The new evidence in this book, documenting the role of al Qaeda in the TWA 800 crash, is equally compelling.

In fact, Yousef's central role as al Qaeda's key operational point man is key to understanding the depth and breadth of planning that went into the 9/11 plot. The most disturbing aspect of the 9/11 Commission's selective review of the evidence is that virtually all of the clues to bin Laden's ever growing threat to America were contained in the files of the U.S. Justice Department—specifically the FBI's New York office and the office of the U.S. Attorney for the Southern District of New York (SDNY), the "offices of origin" for the Yousef-bin Laden cases.

One reason that the 9/11 Commission came up with only half of the story is that the staff chose to limit its investigation to the late 1990s—by which time the plot was fully operational. A true appreciation of the continuing danger that al Qaeda poses to America requires an examination of the roots of bin Laden's first brick-and-mortar al Qaeda cell, the members of which were fully known to the FBI more than fifteen years ago.

The Calverton Shooters

Over four successive weekends in July 1989, the FBI's Special Operations Group (SOG) followed groups of Middle Eastern men (known as ME's) from the Alkifah Center at the Al Farooq mosque in Brooklyn, to a rifle range in Calverton, Long Island. Firing AK-47s and other semi-automatic weapons, the Islamic cell members were captured in dozens of photographs by the FBI surveillance team.

Of the men caught on film by the SOG, one—El Sayyid Nosair—went on to kill right-wing Jewish Rabbi Meier Kahane in 1990.[14] Three of them—Mahmud Abouhalima, Mohammed Salameh, and Nidal Ayyad—went on to serve as Yousef's co-conspirators in the 1993 WTC bombing.[15] Another—Clement Rodney Hampton-El—was convicted in the 1993 Day of Terror plot to blow up the bridges and tunnels around Manhattan.[16] And their leader—Ali Mohammed, an ex-Egyptian Army officer then serving as a Sergeant in the U.S. Special Forces—went on to train Osama bin Laden's personal bodyguard in Khost, Afghanistan in 1996, and personally surveilled the U.S. Embassies in Tanzania and Kenya, taking the pictures bin Laden himself used to pinpoint the locations of truck bombs that killed 234 people and wounded more than four thousand[17] in 1998.[18]

The FBI's thick file of Calverton surveillance photos offered a kind of class yearbook of the terrorist cell that would threaten the United States throughout the next twelve years. For unknown reasons, however, by the fall of 1989 the Bureau terminated its surveillance of the MEs from the Al Farooq mosque.

Had they pressed further, the FBI might have discovered that the Alkifah Center, on the first floor of the mosque, was the New York outpost for the Services Office network, also known as MAK. Since the mid-1980s, the MAK had been taking in millions of dollars a year as the primary funding conduit for Osama bin Laden's "Afghan Arabs," fighting the Soviets in Afghanistan. By 1988, the Saudi billionaire had morphed the worldwide network of MAK locations into a burgeoning new terror network. He called this network "the base" or "the foundation," translated in Arabic as "al Qaeda."[19] Once the Soviet "infidels" had been defeated, the network was rededicated to a violent worldwide "jihad," or holy war. And this time its bloodthirsty efforts were directed at Western nations—particularly the United States, which was clearly a growing, and unwelcome, presence in the Islamic world.

Egyptian radicals like Dr. Ayman al Zawahiri, Mohammed Atef,[20] and the blind cleric Sheikh Omar Abdel Rahman had supported bin Laden in a coup that usurped the MAK cash machine from Abdullah Azzam,[21] a Palestinian scholar who was killed along with his two sons in November of 1989.[22] Bin Laden, who feigned grief at their deaths, was a prime suspect in the murders, and once the Azzams were eliminated, he turned the worldwide string of brick and mortar MAK centers into local al Qaeda outposts. The Alkifah Center was the site of another coup d'etat that took place after bin Laden sent Sheikh Rahman to the United States, where he slipped past a Watch List in 1990.[23]

Picked up by Mahmud Abouhalima, the six-foot-two, red-headed Egyptian and Calverton shooter who became his driver, the Sheikh soon began preaching at three New York area mosques and laying plans to topple Mustafa Shalabi, the Egyptian who ran the Alkifah Center.

Rabbi Kahane: The First Al Qaeda Victim on American Soil

On the night of November 5, 1990, El Sayyid Nosair*—another Egyptian and another of the MEs photographed by the FBI at Calverton—rushed into the Marriott Hotel on Lexington Avenue and gunned down Rabbi Meier Kahane,[24] whose Jewish Defense League had advocated the removal of all Arabs from territories occupied by Israel.

After being wounded in a shootout following the murder, Nosair was arrested. When the FBI and NYPD detectives raided his home in New Jersey, they found Abouhalima and Salameh, two more Calverton shooters, and seized them as material witnesses. That night the Feds also confiscated 47 boxes of evidence—a virtual treasure trove that included bomb receipts, pictures of the World Trade Center, and tapes of the blind Sheikh threatening the "high world buildings" and "edifices of capitalism." More extraordinary, they found top secret manuals from the Army's Special

* Pictures of many of the al Qaeda cell members and leaders, including bin Laden, al Zawahiri, Atef, Rahman, and Abouhalima, can be found on p. 264.

Warfare Center at Fort Bragg, North Carolina, and secret memos linked to the Joint Chiefs of Staff, along with 1,440 rounds of ammunition.[25]

"If there was ever evidence of an international terrorism conspiracy," says retired U.S. Postal Inspector Frank Gonzalez, "this was it."[26] But after NYPD chief of detectives Joseph Borelli, who was eager to avoid a "show trial," declared the Kahane hit a "lone gunman" shooting, the evidence got passed back and forth between the Feds and the cops—to the point where it became effectively inadmissible.[27]

Because of a shortage of Arabic translators, the Sheikh's threatening tapes weren't transcribed into English until after the Trade Center bombing in 1993. Meanwhile, Abouhalima and Salameh, who became two of Yousef's closest cohorts in the bomb plot, were released that night in 1990, and bin Laden's New York cell continued to prosper.

The NYPD soon had another run in with Abouhalima. Throughout the fall of 1990, Sheikh Rahman began openly denouncing Mustafa Shalabi, the Egyptian who ran the Alkifah Center. Shalabi had once been so important to bin Laden that the Saudi billionaire had used him when he made his move from the Sudan to Afghanistan.[28] But al Qaeda had designs on his Center. Fearing that his life was in danger, Shalabi made urgent plans to flee to Egypt in late February of 1991, on the eve of the Gulf War. But before he could get away, he was murdered.

At the time of Shalabi's homicide, more than $100,000 in Alkifah funds were missing; Shalabi's bludgeoned body was found with two red hairs in his hand.[29] Abouhalima (also known as "the Red") identified the slain Egyptian for the NYPD, but despite the forensic evidence he was never charged in connection with the murder; nor was the Sheikh, who had a perfect motive for the takeover of the MAK center. It was the second al Qaeda-related homicide in New York City in three months.

The FBI's Mole Inside Al Qaeda's New York Cell

The first major signal to the Feds that there was a link between what they perceived to be a "loosely organized" group of Arabs and bin Laden's nascent terror network occurred shortly thereafter in 1991, when Nosair's cousin, Mohammed El-Gabrowny, traveled to Afghanistan and got a $20,000 contribution from Osama bin Laden for El Sayyid's defense fund. Although the Bureau's two principal Nosair watchers, Detective Lou Napoli and Special Agent John Anticev, were aware of his ties to the al Gamma'a Islamiyah (IG), an Egyptian al Qaeda affiliate run by the blind Sheikh, no one at the NYPD-FBI Joint Terrorist Task Force seemed aware enough of the emerging threat to put that bin Laden dot on the chart.

The impact of this early warning cannot be understated. One of the major revelations of the 9/11 Commission in its final report was that a 1998 Presidential Daily briefing, revealed a plot by the IG to hijack a plane to free Sheikh Rahman, who was then in U.S. custody. There was proof in the FBI files *seven years earlier* of the importance of the blind cleric to bin Laden's worldwide jihad.[30] But senior management in the Bureau's New York office simply failed to connect the dots.

Back in 1991 the next, almost unbelievable, series of missteps came after Nancy Floyd,* a tough young Texas-born "street agent" in the FBI's New York counter-terrorism squad, stumbled onto an incredible find. While working Russian foreign counter-intelligence, Floyd went to a midtown Manhattan hotel and came across Emad Salem, an ex-Egyptian army officer with an intelligence background who had become a naturalized U.S. citizen.

Employed as a guard and odd-jobs man at the hotel, Salem began doing small favors for Floyd, tipping her to INS violators and even identifying an Egyptian UN worker who was part of an illegal CIA operation.[31]

Having proved his reliability, Salem told Floyd about Sheikh Rahman, a man he characterized as "much more dangerous than the worst KGB hood."[32] Salem also described the Masjid al Salaam (Mosque of Peace) where the Sheikh held forth in nearby Jersey City as "a nest of vipers."

Intrigued, Floyd convinced her bosses on the Soviet side of the FBI's New York office to recruit Salem as an "asset." For five hundred dollars a week plus expenses, the Egyptian, who was anxious to get back into intel

*See pictures, p. 267.

gathering, was given six weeks to see if he could infiltrate the Sheikh's cell. "He did it in two days," said an FBI source who worked the office back then.[33]

Soon Salem had become so close to the virulently anti-American Sheikh that the cleric trusted him enough to order a *fatwa,* or death warrant, on Egyptian president Hosni Mubarak.[34] In media photographs of Sheikh Rahman, Salem was seen at his side, acting as his interpreter and bodyguard.

The Egyptian asset was also summoned to the state prison at Attica, where El Sayyid Nosair had been locked up for the Kahane shooting. Demanding vengeance against the judge who sentenced him, and a pro-Israeli assemblyman from Brooklyn, Nosair exhorted Salem to take the lead in a plot to bomb a series of twelve "Jewish locations," including New York's diamond district.[35] Salem brought the news to Floyd because Napoli and Anticev, the two JTTF "control agents" to whom he'd been assigned, were often unavailable.

Having an asset like Salem inside the Sheikh's cell was an extraordinary intelligence opportunity for the FBI. Sheikh Rahman was beloved of bin Laden. The Saudi billionaire later adopted two of the Sheikh's sons, and he was fiercely loyal to the Sheikh for siding with him along with Egyptians al Zawahiri and Atef in his elimination of Abdullah Azzam. Now, based on his dangerous undercover dealings with the Sheikh, Salem was predicting that there would be "blood in this city" if the cleric wasn't stopped.

But the FBI soon squandered the chance when Carson Dunbar, an FBI administrative special agent in charge (ASAC), was assigned to run the New York office's terrorism unit. An ex-New Jersey state trooper with little experience in the field and no background in counter-terrorism, Dunbar quickly proceeded to alienate Salem, demanding that he take multiple polygraphs, even after he had proven his reliability. Even worse, ASAC Dunbar insisted that Salem wear a wire, and that he agree to testify in open court.[36] But Salem, who had family in Egypt, was extremely fearful of the Sheikh. From the start of his arrangement with the Bureau he had insisted on serving as a pure intelligence asset, with the understanding that he stay permanently under cover and never be identified.

Once Salem had offered evidence of a potential crime like the bombing plot, it was up to the Bureau to perform the necessary surveillance to establish probable cause for an indictment. But Dunbar changed the rules of engagement.[37]

Len Predtechenskis, a decorated former agent who was Floyd's mentor, recalls their frustration at Dunbar's decision. "Nancy believed, as I did, that we could have kept Salem under," he says. "There were various options we could have pursued. We could have recruited another agent to go in to corroborate what was going on and later testify. We could have done surveillance. But there was no effort by management to find anybody else or get wiretap warrants. Management had already made up its mind. Salem had to testify or he was out."[38]

So Salem politely withdrew. Given three months to find another paying job, he met in the fall of 1992 with agent Floyd to receive his last pay envelope. By that time, having lost the trusted Salem, Sheikh Rahman contacted Pakistan, and Ramzi Yousef was sent to New York City. Suddenly a bona fide bomb maker had entered the heart of the Sheikh's cell. And soon the ambitious Yousef had escalated Nosair's "twelve Jewish locations" plot into the World Trade Center bombing conspiracy.

"Don't Call Me When the Bombs Go Off"

The master bomb maker had slipped into Kennedy Airport on the night of September 1, 1992, in a first-class seat beside Mohammed Ajaj, a Palestinian who had been a member of the al-Fatah terrorist group. Both of them carried papers linking them to Al-Bunyan, a cultural center in Tucson controlled by a charity directly linked to al Qaeda and Wadih El-Hage, a Lebanese-born, naturalized U.S. citizen who became Osama bin Laden's personal secretary in Sudan and was later indicted as a co-conspirator in the African embassy bombings.

El-Hage, a key link to the Phoenix, Arizona, end of the 9/11 plot, had been in New York earlier to visit El Sayyid Nosair in jail.* All of these

*See pictures, pp. 264–5.

connections between Yousef and bin Laden were ignored in the 9/11 Commission's Staff Statement, which called the question of Yousef's al Qaeda membership "a matter of debate."[39]

Back in September 1991, however, it was Ajaj's role to serve as a "smokescreen" to facilitate Yousef's entry. As he and Yousef arrived at JFK and went to separate INS stalls, the Palestinian was carrying a suitcase full of bomb manuals and a videotape of suicide car bombers, along with multiple immigration documents, including a passport with his picture crudely pasted over that of a Swedish national.[40] While agents arrested him, Ajaj caused such an uproar that Yousef, claiming political asylum, was given a hearing date and welcomed into the country.

Now, funded with wire transfers from bin Laden, his uncle Khalid Shaikh Mohammed, and other al Qaeda assets, Yousef set up shop in Jersey City. There he worked with three of the Calverton shooters, who had been known to the FBI since the summer of 1989: Abouhalima, the redheaded Egyptian; Salameh, the diminutive Palestinian; and Nidal Ayyad, a Kuwaiti Rutgers grad who worked in a Garden State chemical company.

Soon Yousef was at work building the 1,500-pound urea nitrate-fuel oil (UNFO) device that his cell would plant beneath the North Tower of the World Trade Center. And before long Salem had caught wind that something was afoot. Though he had no specifics, Salem told Special Agent Floyd what he knew during his final meeting with her at a Subway sandwich shop near the FBI offices at 26 Federal Plaza. During the brief meeting, Salem pleaded with Floyd to ask fellow investigators Napoli and Anticev to "follow Mahmud and Mohammed."* Abouhalima and Salameh, he felt sure, would lead the FBI to whatever was in the works. His last words to her were chilling: If the Bureau didn't mount the surveillance, Salem warned, Floyd shouldn't bother to call him "when the bombs go off."

But mistrusted by Carson Dunbar, Floyd had been branded "a bitch" by the ASAC's subordinate, who wanted her off Salem's case.[41] With counter-terrorism experience well below his pay grade, Dunbar was unwilling to back any surveillance based on the Egyptian's advice,[42] So Napoli and Anticev never crossed the Hudson River to look for the Red,

* See pictures, pp. 264–5.

even though Abouhalima was making daily trips to Yousef's bomb factory on Pamrapo Avenue in Jersey City as he built the UNFO device.

Repeated Lost Clues to the Bombing Plot

Abouhalima had been under surveillance by Napoli and Anticev for years. They had earlier obtained enough probable cause for a warrant to search his apartment; and he'd boasted about leading them on "wild goose" car chases. If Abouhalima's phone records had simply been monitored by the Bureau pursuant to Title III warrants, they could have seen traffic to and from a pay phone outside Yousef's bomb factory, including four calls from the World Trade Center made during a surveillance trip.[43]

Mohammed Salameh was even more visible in the fall of 1992 and early months of 1993, when three separate traffic accidents put him in contact with the police. Yousef was injured in one crash. Hospitalized under his own name, he audaciously ordered chemicals from his hospital room via an illicit phone credit card.

Another astonishing event occurred in December 1992, when Federal Judge Reena Raggi defied the efforts of federal prosecutors and ordered Ajaj's bomb manuals returned to him. One of Ajaj's books was marked with an Arabic word that the Feds mistakenly translated "the basic rule." In fact, the phrase was "al Qaeda"—"the base."[44] We'll examine that decision in depth later, as Judge Raggi comes to play a role in another significant aspect of the story.

In the meantime, it should be noted that the Feds blew another opportunity to capture Yousef when he regularly engaged in phone calls with Ajaj who was then serving time in federal prison. The Palestinian would converse with Yousef in code, via a third-party call patched through his uncle's hamburger restaurant in Texas and forwarded to the pay phone outside the Pamrapo bomb factory. Once again, the Feds apparently failed to monitor the calls or translate them until after the Trade Center bombing.

"The only way to explain this Keystone Cops level of negligence," says ex-federal agent Gonzalez, who worked later on the Yousef defense team, "is to understand that the FBI and Justice saw these guys as a kind of pick-up team of ragheads. They didn't seem to have a clue at the time that this was a tightly coordinated cell, with bin Laden calling the shots."[45] The irony is that in 2004, the 9/11 Commission had the same limited view of Yousef's 1993 cell.

Warning That Al Qaeda Would Return to Attack the WTC

On February 26, 1993, Yousef's device went off on the B-2 level beneath the North Tower of the Trade Center, killing six and wounding more than a thousand people. Yousef fled New York via the first-class lounge of Pakistani Airlines at JFK. But before he left, evidence suggests that he contacted Nidal Ayyad,* another of the Calverton shooters who had been on the FBI's radar since 1989. Jealous that the Feds were blaming his work on "Serbian" terrorists, the audacious Yousef is believed to have called Ayyad and ordered him to type an addendum onto the letter that took credit for the blast in the name of what he called "The Fifth Battalion of the Liberation Army."

If the Feds ever needed a signal of what was to come, they found it days later when they searched Ayyad's computer. Originally hoping that the bomb would send the North Tower crashing into the South Tower and creating a "Hiroshima-like event," Yousef admonished Ayyad to add a new close to the letter. "Our calculations were not very accurate this time," the warning said, "However we promise you that the next time it will be very precise and the Trade Center will be one of our targets."[46]

It took Ramzi Yousef and his uncle Khalid Shaikh Mohammed a little more than eight and a half years to make good on that threat.

* See picture, p. 265.

3

THE SUICIDE-HIJACK PLOT

After fleeing New York on the night of the first World Trade Center attack in February 26, 1993, Yousef ultimately connected with his uncle Khalid Shaikh Mohammed in Pakistan. After a series of meetings, the two began preparations that would escalate Osama bin Laden's al Qaeda *jihad* to previously unthinkable new levels of violence.

One of the most devastating acts of Islamic terror perpetrated against the west had been the downing of Pan Am Flight 103 on the night of December 21, 1988. When a Toshiba boom box packed with ten to fourteen ounces of Semtex (a Czech version of C-4 plastic explosive) detonated over Lockerbie, Scotland, it ripped open the Boeing 747, killing 259 passengers and crew on board. Another eleven people died on the ground.

Since then, airports around the world had initiated heightened screening procedures for airline passengers and baggage. Many international airports began using devices designed to detect the presence of explosives in carry-on luggage, so in consultation with his uncle Khalid Shaikh Mohammed during the summer of 1994, Yousef was determined to create a device that would be "undetectable" to conventional airport screeners.

For their staging point they selected the Philippines, an English-speaking third world country with a large Islamic population and notoriously lax airport security. Yousef knew the country well, having visited Basilan in the Southern Philippines in the early 1990s to train members of Abdurajak Janjalani's Abu Sayyaf Group (ASG), the principal al Qaeda affiliate in southern Asia.

But first the devices had to be designed and tested. So in the early fall of 1994, his eyes partly damaged from the premature explosion of a bomb meant to kill Pakistani prime minister Benazir Bhutto, Yousef and Abdul Hakim Murad,* a fellow Balucistani he had known since 1983, retreated to the slums of Lahore for eighteen days of what Ramzi called "chocolate training."[1]

There, in relative obscurity, the bomb maker would test the timing device on what was soon to become a revolutionary "bomb trigger," powered by a Casio DBC-61 databank watch. With a C106D resistor soldered inside—a component federal bomb experts would later describe as Yousef's "signature"[2]—the Casio timer would be connected to a circuit that included two nine-volt batteries to increase power, a broken bulb "initiator" to cause a spark (once the timer counted down to the moment of ignition), and a high explosive, which Yousef determined should be nitrocellulose or "gun cotton."† Easily created by adding cotton balls to a solution of diluted nitroglycerine, a small quantity of nitrocellulose wasn't powerful enough to bring down a plane on its own. Strategically placed near a 747's fuel tank, however, such a device could detonate the jet fuel and transform the aircraft into a flying bomb.

As Morris Busby, the State Department's former coordinator for counterterrorism, later said, intelligence experts would come to view Yousef with a combination of contempt and respect: "I don't want to glorify him, but he is something of a genius bomber."[3]

* See picture, p. 264.

† See picture, p. 300.

The Undetectable Bomb

By the late fall of 1994 Yousef was ready to perform a wet test of the device he hoped would become the center piece in a plot designed to blow up as many as a dozen jumbo jets carrying U.S. tourists home from Asia. To identify the mission, Yousef chose a word he had learned while reportedly advising Bosnian al Qaeda members: "Bojinka," the Serbo-Croatian word for "big noise." Continuing to refer to his cell as the "Liberation Army," Yousef traveled to Manila in December and set up a bomb factory in Room 603 of the Dona Josefa apartment building.

The cell would ultimately consist of three other principals: his old friend Abdul Hakim Murad, a commercial pilot trained in four U.S. flight schools;[4] Yousef's uncle Khalid Shaikh Mohammed, by then a senior al Qaeda operational commander; and Wali Khan Amin Shah, aka "Osama (or Usama) Osmorai,"[5] an Uzbeki veteran of the Afghan war and a close friend of Osama bin Laden.[6]

According to formerly classified documents from the Philippine National Police (PNP), the cell was financed by bin Laden's brother-in-law Mohammed Jamal Khalifa[7] through a Malaysian-based "cut out" or front company called Konsonjaya. Yousef and his uncle used a series of Filipina bar girls (Carol Santiago, Aminda Custodio, and Rose Mosquera) to help them open the bank accounts and obtain the cell phones they would need for the plot.

A once-secret chart,* obtained by the author from the PNP, shows how the money was funneled from bin Laden through Khalifa to the Abu Sayyaf Group (ASG), al Qaeda's primary Philippines affiliate,[8] and on to the cell comprising Wali Khan, Yousef (aka Adam Ali), Murad, and Yousef's uncle Khalid Shaikh Mohammed (aka Salem Ali).†

One of the co-conspirators in the plot, who sat on the board of Konsonjaya with Shah, was an Indonesian cleric named Riduan Ismuddin, aka "Hambali." To illustrate how tightly connected this Yousef-KSM cell was to Osama bin Laden and al Qaeda, consider that Hambali was one of the attendees of the now infamous January 2000 9/11 planning session in

* See flow chart, p. 309.

† See pictures, pp. 264–5.

Kuala Lumpur, Malaysia, monitored by the CIA. Hambali* was later tied directly to al Qaeda's November 2002 Bali nightclub bombing, which left 180 dead.[9]

As we'll see in Chapter 17, one of the 9/11 Commission's most astonishing lapses was the staff's dismissal of this evidence—backed by PNP files—which proved that Osama bin Laden himself bankrolled the Yousef-KSM-Murad-Shah Philippines cell; evidence that in two months working out of the Dona Josefa apartments, Yousef spawned not only the Bojinka and Pope plots, but the 9/11 attacks.[10]

The first target was the pontiff, who was scheduled to visit Manila on January 12, 1995. To bring off the assassination, Yousef designing a series of remotely detonated pipe bombs to be planted along the Pope's parade route. Since the onlookers along the route would be packed twenty-five to fifty deep, the scheme promised to kill thousands.[11]

Later, an interrogation of Yousef's partner Abdul Hakim Murad would reveal the details of the suicide-airliner hijacking attacks ultimately carried out on 9/11. Before that plot could be realized, however, Yousef designed the Bojinka plan. Eventually described by Federal prosecutors as "48 Hours of Terror," the plot involved the smuggling of Yousef's Casio "bomb triggers" aboard American jumbo jets leaving Asia. This was not a suicide scenario. The plan was to use aircraft flying in two stages. Each of the conspirators would board on the first leg of each flight, assemble the devices, plant them and exit the aircraft only to repeat the process three more times over two days.

Designed to avoid detection by airport screeners, the devices would be placed under seats where life jackets were stored, and be timed to detonate on the second leg long after the bombers had left the planes. It was Yousef's intent to place the bomb triggers over the center wing fuel tanks of the Boeing 747-100s so that the Casio-nitro devices would act as improvised "blasting caps"to ignite the jet fuel and blow the tanks.

Later refined by Yousef, the plan ultimately called for the plotters to board eleven flights over a forty-eight-hour period.† Yousef, who had as

* See picture, p. 265.

† Beyond Yousef, Murad, Shah, and KSM the identify of the fifth Bojinka co-conspirator was never revealed publicly.

many as a dozen aliases himself, created clandestine code names for the plotters: Obaid, Zyed, Markoa, Majos, and Mirqas.* He created intricate flight schedules for the plot, which would require precise timing, and stored all of the details on his Toshiba laptop. The bomb maker, who always showed a flair for the dramatic, used the same laptop to create elaborate fake IDs for the plotters.

In a part of the Bojinka schedule later recovered by the PNP, Yousef put his own TOP SECRET designation on the time schedule for the elaborate two-day plot.

In one case, shown on page 301, the plotter code-named "Obaid" would board United Flight 806 in Singapore (SIN). In the first leg of the flight, bound for Hong Kong (HNG), he would assemble and plant the device. Obaid would then deplane in Hong Kong as the flight jetted east, bound for San Francisco (SFO). The device was timed to go off ten hours later, when the flight was still above the Pacific. The plotter would then board another two-leg U.S. bound flight in Hong Kong and repeat the rotation. If fully executed, Federal prosecutors later determined, the plot would have killed up to four thousand Americans.

The Wet Test

Now, on December 1, 1994, with his Manila cell in place, Yousef ordered the first test of his "undetectable bomb." He sent Wali Khan Amin Shah to set a small explosive charge, using the Casio timer, under a seat in Manila's Greenbelt theater. At 10:30 P.M. setting the databank watch five minutes ahead and placing the device under the ninth seat in the fourth row from the movie screen, Shah quickly snuck back to his motorcycle in the street outside the theater and waited.

At exactly 10:35 P.M. he heard the blast, and saw terrified patrons rushing from the theater covered in blood and debris. The test was a success.

Ten days later, on December 11, Yousef decided to test the limits of airport security at Benino Aquino International Airport in Manila.

*A reputed fifth plotter was never identified by the Feds.

Assuming the identify of an Italian National named Arnaldo Forlani—a variation on the name of an Italian diplomat he had recovered from the atlas in the Toshiba's Windows software—Yousef purchased a one-way ticket on a Philippine Airlines 747-100 on the first leg of a two-leg flight from Manila to Cebu, a small city in the Southern Philippines.

Yousef succeeded in defeating the security checkpoint, in part by smuggling one nine-volt battery in the heel of each shoe.* When his Casio watch set off the alarm, he proffered a religious medal around his neck, and the security screeners, eyeing the Italian passport, took him to be a Catholic. Hidden away in his shaving kit, the diluted nitro (in a contact-lens cleaner bottle), his Toshiba carrying case (wires and all), and the penlight he would use for the bulb initiator failed to arouse suspicion.

On board after takeoff, Yousef assembled the device in a series of quick, rehearsed moves. Working in the back of the plane rather than the lavatory, where his extended stay might have aroused suspicion, the "evil genius" snapped the batteries onto the circuit, plugged one end into a female connector protruding from the C106D resister inside the watch, and connected the other end of the circuit to the E72332 flashlight bulb, which he broke, exposing a "pig's tail" filament. He pushed the cotton balls into the nitro (filling the contact lens cleaning bottle), carefully inserted the broken bulb initiator, taped on the cap, and then set the Casio four and one half hours ahead, to a time when he would be on his way back to Manila while PAL Flight 434 jetted across the South China Sea toward Japan.

Zipping the contents of the "bomb trigger" back into the shaving kit, Yousef slipped into seat 26K.† Then, after looking around to make sure he wasn't spotted by other passengers on the early-morning flight, Yousef hid the improvised explosive device in the pouch where a lifejacket was stored below the seat.

After touching down at 6:25 A.M., he left the plane in Cebu City just as Haruki Ikegami, a twenty-four-year-old engineer, was being given the seat assignment 26K by gate personnel. Eager to create an alibi, the wily Yousef then paid cash for a flight that landed him back in Manila by late morning.

*After determining via a CNN report that the metal detection system began an inch off the ground.

†He was under the impression that the center wing fuel tank of the 747 began at row 26.

Months later, investigators concluded that he must have slipped up the back stairs of the Dona Josefa, avoiding the security guard in the lobby, and emerged later that day around 1:00 P.M. as the engineer Naji Haddad.

By 10:44 A.M., however, PAL 434 was at 33,000 feet, approaching point Mike Delta near Okinawa, when the Casio timer counted down, sending a low-voltage surge across the circuit. Enhanced by the nine-volt boost from the batteries, it created a charge that caused a spark in the filament of the broken bulb, which in turn detonated the gun cotton. The blast blew a hole in the 747's floor and immediately killed Ikegami, whose body was severed by the explosion. The pilot, Captain Edwardo Reyes, signaled a mayday message from the cockpit as he tried to hold the huge jumbo jet steady; and his co-pilot contacted air traffic controllers. Reyes momentarily lost the ability to steer the plane before finally managing to turn the 747 and making an emergency landing at Naha Airport in Okinawa.

It had been Yousef's intent to blow the center fuel tank, exploding the 747 with 293 passengers and crew members on board—not because he had any particular grudge against the Philippine government, but because he needed to wet-test his device. Once he heard the press reports of the emergency landing, he realized that he had not placed the device far enough forward—a mistake he hoped to correct when the full Bojinka mission was operational.

As he watched the CNN footage of the emergency landing and eyed the damage to the passenger cabin* over drinks with Khalid Shaikh Mohammed, Yousef vowed that the other Bojinka "bomb triggers" would be placed farther forward, ensuring that the fuel tanks of the 747s would blow. "Imagine Bojinka times twelve," he said to his uncle. "As Allah wills it," Mohammed smiled, "so it will be done."

Col. Fariscal's "Act of God"

"They might have pulled it all off. The Holy Father might have been killed and all those planes could have gone down, except for God's will." That's how PNP retired police Colonel Aida Fariscal remembered the night of

* See picture, p. 270.

January 6, 1995. Working in his underwear inside Room 603, which was littered with chemicals, pipe bombs in various stages of assembly, a map of the pontiff's parade route, and a series of disguises including priest's cassocks, Yousef was in a kitchen area showing Murad how to mix two chlorine-based chemicals when the addition of sugar to the mixture suddenly caused the compound to ignite.

Neighbors screamed at the explosion; clouds of smoke billowed from the small apartment, the fire department was called, and the two bomb makers rushed from the room, pulling on their pants as they fled.

Eventually, the firefighters extinguished what amounted to a small sink fire and Yousef assembled with Murad and Shah at a nearby 7-Eleven to wait until the authorities had left. He instructed Murad to return to the Dona Josefa to retrieve his coveted Toshiba laptop, but when Murad got there he was promptly seized by Fariscal* and two other PNP officers. Once she entered the bomb factory and saw evidence of the threat to the Pope, she sounded an alarm and Murad was taken into custody.

But not before he ran from the arresting officers. When one of them fired a shot in his direction, Murad tripped. Finally subdued, Murad tried to offer Fariscal a $2,000 bribe as his hands were being tied behind him. "That's when I knew," she later recalled, "that this wasn't some tiny plot with local extremists. With money like that, and the bombs and the pictures of the Holy Pope, I knew that this was something much bigger."

No one in the PNP had any idea that this accidental fire had led them to the lair of Ramzi Yousef, the world's most wanted terrorist. Minutes after witnessing Murad's arrest, Yousef contacted Khalid Shaikh Mohammed, and the two of them left the Philippines for Pakistan. Wali Khan Amin Shah was arrested by the PNP days later, only to escape.

That night Yousef's old friend Murad was the only one of the Bojinka plotters to be seized. Some time after midnight, he was back-cuffed, blindfolded, and taken across Manila's Pasay River to Camp Crame, home of the PNP's Intelligence Group.

Initially, the evidence suggests that he was harshly treated, perhaps even tortured, forced to ingest massive quantities of water. Within days,

* See picture, p. 267.

however, he was handed over to Col. Rodolfo Mendoza,* the PNP's leading specialist on Islamic terror groups.

Proof That the 9/11 Plot Was in Motion

The soft-spoken Mendoza was the antithesis of the cattle prod specialist who uses torture as a means of opening up prisoners. "My approach was guile," he told me in an interview conducted in his offices at Pagasinan in the Northern Philippines. Using a combination of food and sight deprivation and false threats to turn Murad over to the CIA or the Mossad, Mendoza soon gained the terrorist's trust.

After admitting his involvement in the first World Trade Center bombing, bragging that he had chosen the Twin Towers as a target himself, Murad ultimately confessed that the leader of his cell was Abdul Basit, a.k.a. Ramzi Yousef.

"We were really shocked," said the Colonel. "At this time Yousef [was] one of the biggest fugitives in the world."

Little by little, as Mendoza peeled back the layers, Murad began to reveal details of the Bojinka plot. Then, on January 19, 1995, he confessed a tentative plan to attack CIA headquarters in Langley, Virginia with a small plane. He took credit for the idea himself.

MURAD: I told Basit that there is a planning, what about we dive to CIA building.

MENDOZA: CIA building?

MURAD: Yes. He told me [illegible]

MENDOZA: And you are willing to die for Allah or for Islamic—

MURAD: Yes.

In a January 20 debriefing memo after that interrogation session, Mendoza described the plan to "dive-crash a commercial aircraft at the CIA

* See picture, p. 267.

Headquarters." He wrote that the concept had come out of Murad's "casual conversation with Abdul Basit [Yousef]" and concluded at the time that there was "no specific plan" for its execution.

The transcript of that session, which showed up as part of the evidence in the 1996 Bojinka trial of Yousef, Murad, and Shah, created a great misconception within the U.S. intelligence community that continues to this day. Not only did the Congressional Joint Inquiry repeat the "small plane into CIA" story, concluding that "the plans to crash a plane into CIA headquarters and to assassinate the Pope were only at the discussion stage,"[12] but the 9/11 Commission staff also gave it credence: "Two of the [Bojinka] perpetrators had also discussed the possibility of flying a small plane into the headquarters of the CIA."[13]

As noted in *1000 Years for Revenge,* the almost casual uncertainty of Mendoza's briefing memo left the public impression that Murad had said little that would suggest a direct connection to the 9/11 attacks. One former intelligence official, quoted in the *Washington Post,* noted that it was a real leap from "stealing a Cessna to commandeering a 767."[14]

But in the course of our conversation, Mendoza revealed that Murad had ultimately confessed to something far more important: That by the time of their conversation the full blown hijack-airliners-suicide plot executed on 9/11 had already been set in motion by Yousef's cell.

"He discussed with me," said Mendoza, "even without me mentioning, that there is really formal training [going on] of suicide bombers. He said that there were other Middle Eastern pilots training and he discussed to me the names and flight training schools they went to. This is in February of 1995."

In the videotaped interview, which I submitted to the 9/11 Commission following my testimony, Mendoza insisted that Murad had told him that "the plan is to hijack airliners." Three targets, he said, were initially mentioned, including the Pentagon, CIA headquarters, and a nuclear facility. The plot would be accomplished by hijacking commercial airliners, not a small plane as was mentioned in the January 20 debriefing memo. Murad himself had boasted to Mendoza about his extensive training in U.S. flight schools.

In an interview with correspondent Maria Ressa of CNN, Rigoberto "Bobby" Tiglao, the spokesman for then-Philippines President Ramos, expanded Murad's confessed list of targets to include the World Trade Center and two of America's tallest buildings, the Sears Tower in Chicago, and the pyramid-shaped Transamerica Tower in San Francisco.[15]

And while other intelligence authorities, including ex-CIA Director James Woolsey, have often confused the Bojinka plot with Yousef's 9/11 scenario,[16] Mendoza insisted that the two plots were distinct. "Murad is talking about a plan that is *separate* from Bojinka," Mendoza said. "Murad is talking about pilot training in the U.S., and Murad is talking about the expertise of Yousef. Not only was this a *parallel* plot to Bojinka, but what is the motive of Murad to have flight training in the U.S., and who are the other pilots met by Murad in the U.S.? That is an enormous question."[17]

As we'll review in Chapter 18, the Staff of the 9/11 Commission has chosen to ignore Mendoza's evidence, alleging that the idea for the 9/11 plot "appears to have originated with KSM."[18] Quoting Khalid Shaikh himself in Staff Statement #16, they assert that the 9/11 plot was conceived by Mohammed, and presented to bin Laden, two years after Murad told Mendoza Yousef had first planned it. While admitting that skyscrapers on the West Coast were originally intended targets along with the CIA, Pentagon, World Trade Center, a nuclear facility, the White House and Congress, Mohammed—who appears to be the Commission's sole source for the assertion—seems to now be distancing himself from his brilliant nephew Yousef.

"One of the questions I would ask," says terrorism researcher Paul Thompson, author of the Center for Cooperative Research's extraordinary timelines of the events surrounding 9/11, "is why would Abdul Hakim Murad, one of Yousef's oldest friends, have attended four U.S. flights schools to get his commercial pilot's license as early as 1992 if the 9/11 plot wasn't even conceived of until four years later? He didn't need a pilot's license to plant a bomb on a plane in the Bojinka plot."

Mendoza agreed. "Murad is a dedicated man," he asserted. "He provided the structural analysis of the World Trade Center. This may be speculation. But when the truck-bombing scheme did not work [in 1993] they decided that the only way to take down a symbolic target like the World

Trade Center is to steal a plane. He went to the United States and trained in flight training schools, and he told us about taking a plane and hitting the CIA, a nuclear facility, and the Pentagon. Murad was talking about pilot training in the United States, and Murad is talking about the expertise of Yousef. He's talking about his [Murad's] involvement in the WTC casing and the intelligence he developed on the structural analysis. What can you conclude?"

Given that the Commission seems to be relying on Khalid Shaikh Mohammed's word to pinpoint the origin of the 9/11 plots, two additional pieces of evidence are worth considering. First: when Murad was being rendered back to the United States following the conclusion of Mendoza's interrogation, he predicted to FBI agents that Yousef had the Twin Towers in his sights: As FBI agents Frank Pellegrino and Thomas Donlon noted in an FBI 302 transcribed on May 11, 1995, "MURAD advised that RAMZI wanted to return to the United States in the future to bomb the World Trade Center a second time."[19]

Three months earlier, after his capture in Islamabad, Pakistan, Yousef was in a Sikorsky helicopter, surrounded by Federal agents, as the chopper approached the brightly lit World Trade Center in Lower Manhattan. Though there are conflicting accounts of precisely what was said, one witness told *Newsweek* that one of the FBI agents lifted Yousef's blindfold and gestured toward the Twin Towers.[20]

"See," he said, "You didn't get them after all." Yousef reportedly eyed him, then fidgeted in his heavy cuffs and snapped back: "Not yet."[21]

The truth of Yousef's true connections to al Qaeda might have remained locked away inside FBI files. But soon he was reunited with Murad, Wali Khan, and Eyad Ismoil, his World Trade Center bombing co-conspirator, when they were all housed on the secure 9th floor tier of the Metropolitan Correctional Center (MCC), the federal jail in Lower Manhattan.

By the early months of 1996, as Yousef prepared to be tried first for the Bojinka plot, another criminal would be assigned to his adjoining cell. This time, however, his neighbor was no Middle Eastern terrorist, but a forty-five-year-old capo of the Colombo family who had been running numbers in Brooklyn and drugs on Staten Island while keeping the books for his father. Ramzi Yousef, the "Mozart of Terror," was coming face to face with Gregory Scarpa Jr., son of the man they called "The Killing Machine."

As the months passed, Yousef would come to trust the younger Scarpa more than any other "infidel." Soon Yousef came to believe in him so much that he revealed his newest plan to Scarpa—one he had designed to ensure a mistrial in the Bojinka case. In this ingenious plan, members of his cell who were still active in New York and abroad would smuggle a PAL 434-style bomb trigger onto a 747, placing it above the center fuel tank, and blow the plane out of the sky.

Determined to see his Bojinka plot fulfilled, Yousef was anointing Greg Scarpa Jr. as his new co-conspirator. But there was one factor he hadn't counted on: the man he'd selected as his new accomplice was the son of the highest-level known mob informant in the history of the FBI. And running numbers wasn't the only thing he'd learned from his father.

4

"PLAN TO BLOW UP A PLANE"

There was an open-air "rec cage" on the roof of the MCC, but inmates who wanted to use it had to submit to a cavity search each day. Unwilling to undergo such an indignity, Ramzi Yousef opted for an hour of calisthenics in the area just outside of his cell.[1] During these sessions he got friendly with his new neighbor Scarpa.

Jr. may have followed his father into the "family business," but as Yousef soon discovered, he seemed to lack the elder Scarpa's cold-blooded willingness to snuff out human life. Still, Yousef needed a method of getting secure messages to Murad and Shah, who were housed on the other side of Greg's cell on the 9th floor tier along with Eyad Ismoil, a Jordanian who had been Yousef's wheelman back in 1993. So a system was devised whereby Ramzi passed Scarpa notes through a hole in the wall between their cells where a bed strut had been removed. These messages, which he called "kites," sometimes turned into elaborate letters.

Scarpa was facing years in prison for his mob activities. But he was also an American, and he knew full well how dangerous Yousef was. The mob generally avoided civilian casualties, but al Qaeda was dedicated to inflicting them in mass numbers. The younger Scarpa had nothing but contempt for the terrorists. But given his father's success as an FBI informant, Scarpa Jr. realized how valuable an inside connection with Yousef could be. Eager for a "downward departure" date from the Feds under rule 5K1.1 of the U.S. sentencing guidelines,[2] Scarpa used his natural con-man abilities to convince Yousef that La Cosa Nostra hated the U.S. government even more than the right-wing militias of the time.[3] Soon he had gained Yousef's confidence and made arrangements to inform on him to the government.

According to Larry Silverman, his attorney at the time, Scarpa rejected a traditional plea deal, which would have resulted in a seventeen-year term, because "the Feds wanted him to plead to the top count in his indictment and this would have meant too many years inside."[4] Instead, Silverman agreed to a deal providing that if Scarpa furnished "substantial assistance" to the FBI, he would be rewarded with an early release.

Meanwhile, concerned that his client wouldn't be able to prove "substantial assistance," Silverman came up with a scenario designed to help the Feds gather hard intelligence on Yousef while potentially thwarting any acts of terror the leads might generate. Working with prosecutors from both the Southern District and Eastern District of New York, he convinced the FBI's New York office to create a system that allowed Yousef to make third-party calls from the phone on the 9 South tier. The calls would be routed to the fictitious "Roma Corporation," a supposed mob operation with offices on Fifth Avenue. Scarpa would tell Yousef that his "family" members at Roma could receive Yousef's calls, then patch him through to anybody he wanted to talk to, in the United States or abroad. Scarpa also raised the possibility with Yousef that "his people"—meaning mafiosi—could get the terrorists passports.

Aware that this could be an intelligence bonanza, the Feds provided Scarpa with a two-inch "spy camera" to use in photographing Yousef's kites, which soon included detailed bomb formulas and specific threats of terrorist acts. They even introduced a female FBI agent, posing as a paralegal named Susan Schwartz, to visit Scarpa periodically and retrieve the film. In addition, they contributed five hundred dollars to Yousef's commissary account at the prison, which allowed him to buy toiletries and other personal items. But the Feds were so wary of Yousef's skills at making improvised bombs that they wouldn't even let him keep mouthwash or toothpaste in his cell.[5]

At Roma Corp., the Feds planned to have Arabic-speaking translators present to monitor all of Yousef's calls and track the outside numbers to potential cell members abroad.

By early May 1996, as the Bojinka trial approached, Scarpa was poised to hand the FBI and federal prosecutors a treasure trove of information, delivered the through the wall directly from Yousef, Osama bin Laden's chief bomb maker himself.

To protect the government later from "spy in the defense camp" allegations, Valerie Caproni, chief of the EDNY's Criminal Division, assured Reena Raggi, the judge on the Scarpa case, that a firewall had been set up to isolate the Southern District prosecutors trying Yousef from this new operation. She pledged that the Assistant U.S. attorneys and FBI agents processing the Scarpa intelligence would be a "security team," and would not share evidence with the AUSAs prosecuting Yousef in the Bojinka case.

As the operation commenced, however, it is unlikely that such a firewall effectively existed. One of the top Feds overseeing the Scarpa pass-through operation was Assistant U.S. Attorney Patrick Fitzgerald, chief of the SDNY's Organized Crime and Terrorism Unit. It was Fitzgerald who also supervised AUSAs Mike Garcia and Dietrich Snell, the two prosecutors on the Bojinka case. Early on, one of Scarpa's most stunning revelations was that Yousef intended to kill Garcia and a federal judge (believed to be Kevin Duffy, who was about to preside at the Bojinka trial).

"It's difficult to believe, given this threat to Garcia's safety and the clear precautions that were later taken, that he was not made aware of the threat or the source of the intel," said an attorney with knowledge of the case.[6]

Furthermore, as we've seen, Judge Raggi was well acquainted with Yousef, having rendered that extraordinary decision to return Mohammed Ajaj's bomb books to the terrorist prior to the World Trade Center attack in December 1992.* It's clear that Raggi knew the kind of violence Yousef

* As it turned out, Ajaj never claimed the manuals and tapes of suicide bombings; they remained at the FBI's New York office.

was capable of—and it's a safe bet that she monitored the Scarpa surveillance closely.

As the Bojinka trial opened on May 25, then, the Feds knew from Scarpa that the bomb maker had prosecutor Garcia and a judge on his hit list. As the days passed in May, though, Yousef spilled the details of an even more shocking terror plot.

In the almost weekly kites from Yousef, which Scarpa dutifully copied verbatim or photographed, he began to amass intelligence on al Qaeda as detailed and compelling as the finest bugs ever installed by the FBI in the apartment above John Gotti's social club in the 1980s.

"It was weapons grade plutonium," said one former FBI agent who examined the material in preparation for this book.[7] "And it was also a blueprint that could have led the Bureau not just to Yousef's role in the crash of TWA Flight 800, but to 9/11 as well."

"How to Smuggle Explosives into an Airplane"

By the spring of 1996, Greg Scarpa was documenting bomb formulas from Yousef that were nearly identical to those used to build the intricate Casio-nitro device he'd used to kill Haruki Ikegami in the wet test aboard PAL 434 on December 11, 1994.*[8]

The hand drawings Greg did as Yousef bragged about the intricacy of his devices were even more precise and revealing than the half-assembled bombs seized in the PNP raid on Room 603 at the Dona Josefa Apartments, or the formulas on Yousef's Toshiba laptop.

On May 19 Scarpa wrote: "Yousef holding the kite, while I read it through the wall (Sunday or Monday) Yousef advised his people to start the bombing of airplanes."[9] At the top of the kite Scarpa noted that "The Gover[nment] knows this info is very accurate because they heard parts of it on the 3-way line. Susan told me."

Further down in this May 19 note Scarpa added a subhead: "How To Smuggle Explosives Into An Airplane." He then proceeded to take down

* See diagram p. 300.

detailed lists of explosives and bomb schematics that could only have come from Yousef, laying out the precise scenario of the Bojinka plot. "All explosive substances of a density higher than 2 kg/l should not be used due to the possibility of detecting them by X-Ray machines," he copied dutifully. "The following explosive substances have a density less than 2 and can easily be smuggled:

- Tetrazene (guanylnitrosoaminoguanyltetrazene)
- Acetone Peroxide
- RDX
- HMTD (hexamethylenetriperoxidediamine)
- DDNP (diazodinitrophenol)
- HMX (cyclotetramethylenetetranitramine)"*

Corroborating Yousef's smuggling method during the PAL 434 wet test, where he hid nine-volt batteries in his shoe heels, Scarpa quoted the bomb maker as advising that "a detonator can be hid in a heel of a shoe."[10]

Describing a figure believed to be Osama bin Laden and code-naming him "Bojinga," Scarpa then wrote: "Yousef told me Bojinga [sic] himself is in Afghanistan. Yousef's messages to him are sent through Yousef's relatives in Iran. He has friends, most of them are in Saudi Arabia & Qatar, he said but some of them ran away when they were mentioned as co-conspirators in Yousef's case." This was a hint that one of Yousef's ongoing cohorts on the outside was none other than his uncle Khalid Shaikh Mohammed. In fact another line in Scarpa's notes strongly suggested it:

"Yousef said one of them is a close relative of his. His friends call him 'Bojinga' because he uses the word a lot—[But] Yousef said he is not the real Bojinga. The real Bojinga is in Afghanistan."

This revelation was significant because it alerted the Feds to a conclusion that the 9/11 Commission would reach, eight years later—that bin Laden's al Qaeda network received aid and assistance from Iran.[11]

*See Appendix II, p. 302.

In another kite entitled "The explosive used in blowing up airplanes," Scarpa included a schematic, drawn at Yousef's instruction, that matched precisely the design of the bomb trigger that Yousef used in his December 1994 PAL 434 "wet test." A comparison between the schematic and a partially assembled Casio-nitro device can be seen in Appendix II on page 300. The construction of the device involved soldering a high-impedance SCR 106D resistor into a Casio DBC-61 databank watch, both elements identical to components that Yousef had used on PAL 434. There were instructions for snapping a nine-volt battery into the circuit for power, and connecting the timing mechanism to the filament of a broken light bulb, which would provide a spark that would ignite a high explosive mixture. In the wet test Yousef used a diluted nitroglycerine solution hidden in a bottle of contact lens cleaner that was stuffed with cotton balls creating the high explosive nitrocellulose, also known as gun cotton. Earlier in the note to Scarpa, he suggested a choice of explosives including RDX, aka Cyclotrimethylene trinitramine, a high explosive widely used by the military.*[12]

Now, in late May 1996, if the Feds ever needed confirmation that Scarpa (a mobster with a high school education) had direct access to the master bomber, this was it. Further, Scarpa's intel mirrored the evidence that AUSA Mike Garcia and his partner Dietrich Snell were about to present to a jury in Yousef's Bojinka case. Most significant, the kites revealed Yousef's pledge to Scarpa that if the trial went against him, his cell would set off "A BOMB" aboard a U.S. commercial airliner in an effort to affect a mistrial. This threat would take on crucial implications in the weeks ahead as the weight of the evidence turned against Yousef and he desperately sought a way out.

Confirmation of the Kites in FBI 302s

There's no doubt that the Feds were aware of the extraordinary intelligence Scarpa was recovering. In addition to processing the film from the spy camera, FBI agents met with Scarpa and his then-attorney, Larry Sil-

*This mention of RDX becomes very important later on because the identical chemical was found with nitroglycerine in the wreckage of TWA Flight 800.

verman, on a weekly basis to interview him as a way of corroborating the evidence. These interviews were memorialized in a series of FBI 302 memos. Consider the following excerpts from a 302 on May 6, 1996:[13]

> YOUSEF told SCARPA I'll teach you how to blow up airplanes and how to make bombs and then you can get the information to your people [meaning Scarpa's people on the outside]. YOUSEF told SCARPA I can show you how to get a bomb on an airplane through a metal detector. YOUSEF told SCARPA he would teach him how to make timing devices. . . . YOUSEF told SCARPA that during the trial they had a plan to blow up a plane and hurt a judge or an attorney so a mistrial will be declared.*

In that same memo, FBI agents uncovered evidence of an al Qaeda cell loyal to Yousef, then active in New York City. This is another revelation being made public here for the first time:

> According to SCARPA, YOUSEF has indicated that he has four people here. SCARPA described these four terrorists already here in the United States. SCARPA advised YOUSEF has not indicated who these four individuals are or if he is in contact with these four people or how he contacts them. SCARPA does not know if YOUSEF receives or sends any messages from contacts overseas.†

Until now, the conventional wisdom—supported by National Security Advisor Condoleezza Rice in her testimony before the 9/11 Commission—was that after the arrest of Yousef and his cohorts in 1995, there was no evidence of any active al Qaeda terror cell operating in New York City. According to Rice, the next hint of any suspicious activity on U.S. soil was revealed in a section of the August 6, 2001, Presidential Daily Briefing (PDB), which described "the recent surveillance of federal buildings in New York."[14]

But in her statement and testimony before the Commission on April 8, Rice repeatedly called the warnings in the PDB "historical" and devoid of any actionable intelligence that would have given the White House a prior warning of the 9/11 attacks. In response to a question posed by Commissioner Bob Kerrey, Rice insisted that "the problem was that for a country that had not been attacked on its territory in a major way in almost 200 years, there were a lot of structural impediments to those kinds of attacks."[15]

*The 302 can be seen on page 299.

† See excerpt from 302 on page 305.

Rice stressed to the Commission that "if we had known that an attack was coming against the United States, that an attack was coming against New York and Washington, we would have moved heaven and earth to stop it." Yet, as the extraordinary Scarpa-Yousef file amply demonstrates, the bomb maker made *repeated* threats of al Qaeda hijackings if he and his cohorts—including Sheikh Omar Abdel Rahman—weren't freed from prison.

The August 6, 2001, PDB characterized as unconfirmed "sensational threat reporting" a 1988 foreign intelligence report that "Bin Laden wanted to hijack a US aircraft to gain the release of 'Blind Shayk' 'Umar 'Abd al-Rahman and other US-held extremists."[16] The final 9/11 Commission report released on July 22, 2004 also revealed an identical threat from a 1998 PDB sent to President Clinton. Yet, as the FBI files prove, Yousef had made the precise threat to Scarpa almost two years earlier. But once again, Rice and other top national security officials seemed unaware of the intelligence.

A Message to Khalid Shaikh Mohammed

By the week of May 6, the Feds had set up the patch-through phone system via Roma Corp., the FBI mob front company, that allowed Yousef to make third party outside calls. An FBI 302 from Thursday, May 9, transcribed the next day, noted that "YOUSEF mentioned that he needs to make a phone call overseas to find out about the airplane situation."[17] Yousef even told Scarpa that he needed the phone prior to ten in the morning or after 9:30 at night, because of the time difference between New York and the Middle East.

On May 9, Scarpa passed a "kite" to Yousef through the hole. It contained the name and address of a "George Smith," the fictitious wiseguy Scarpa led Yousef to believe would take care of his third party calls from the MCC through Roma Corp. The FBI even furnished Yousef with a fax number for his use in passing information to the Middle East.*

*See picture, p. 301.

Scarpa's note advised Yousef on the protocol for making the outside calls. He was to dial the phone number of the Fifth Avenue address and say that "Ronnie" was calling and would like to make a call. The FBI's team of phony mobsters would then patch through Yousef's call as they monitored it in real time.[18]

Naturally, there was a danger for the Feds. They knew that Yousef had spoken in code and used a similar third-party patching method in 1992 as he built the 1,500-pound World Trade Center bomb, using a Jersey City pay phone to speak with his co-conspirator Mohammed Ajaj, then in Federal prison. The calls were patched between Yousef and Ajaj via a burger restaurant in Texas. Now the Feds were doing essentially the same thing: allowing the bomb maker to speak to al Qaeda cell members outside of the prison, with the hope of listening in.

But as it turned out, rather than speaking Arabic, which the Feds were prepared for, Yousef spoke in Urdu and Baluchi, two of the six languages he used. And, as Scarpa reported later, the bomb maker was apparently able to get key information out to his conspirators before he could be stopped.[19]

One of the people Yousef spoke to via the Feds' phone link was his uncle Khalid Shaikh Mohammed, the fourth Bojinka co-conspirator, who was still at large at the time. In 1996, KSM, as he was known to the Feds, was hiding out in Doha, the capital of Qatar on the Persian Gulf. FBI documents would later reveal that Qatari officials believed Mohammed was constructing an explosive device at the time.[20] The Joint Congressional Inquiry concluded that a "link to KSM was made when Yousef . . . made a call from detention to Qatar and asked to speak with 'Khalid.'"[21] The number Yousef called, as it turned out, was similar to one found by the Philippines National Police when it raided Yousef's Doña Josefa bomb factory and seized his Toshiba laptop.[22]

An FBI 302 dated May 13, 1996, shows that the Feds were so anxious to facilitate Yousef's outside calls that one female counselor at the MCC asked him if he needed a phone, and offered to let him use the one in her office.[23] But at the time the counselors' phones were reportedly down, so in a May 16, 1996, 302 it was noted that a Lieutenant Desman (of the Bureau of Prisons) was "working on the phones to make it easier [for Yousef] to call out."[24]

By May 28, Yousef told Scarpa that he wanted his "people to call in three bomb threats to UNITED AIRLINES international flights" in the same week. "The purpose," according to Scarpa, was to "cause disturbance and fear among people flying airplanes" and to "cause financial problems for the airline."[25]

That same 302 alerted the Feds to the fact that Yousef was willing to set up a meeting "between one of [Yousef's] people and one of SCARPA's people on the outside." Considering what we now know—that Yousef's uncle Khalid Shaikh Mohammed was actively working at the time on the hijack-airliners-suicide plot that resulted in the 9/11 attacks—this was an extraordinary intelligence opportunity for the Feds.

In return for setting up the meeting between his al Qaeda cohorts and the ersatz Mafiosi, all Yousef was asking was $2,500.

By June 11, as the Bojinka trial played out and the evidence was turning against Yousef (who was representing himself), Scarpa reported yet another airline bomb threat. On June 11, Special Agents Charles Provine and Pat White wrote that "SCARPA recently asked [Abdul Hakim] MURAD about the alleged plot to blow up a United States airliner." Murad replied, "RAMZI's waiting to hear if BOJENKA got the message." (Again, Yousef was using the name BOJINKA—or "BOJENKA," as the Feds spelled it—to refer to Osama bin Laden.)

Three days later, the intelligence from Scarpa grew even more urgent. On June 13, 1996, Special Agent White got an unsolicited call from Scarpa. It came while White was at his office inside the Joint Terrorist Task Force at the FBI's New York office, right across Foley Square from the MCC.

Scarpa told him that Yousef was anxious for "his group" to meet with Scarpa's people, but he wanted the sit down to take place "where his (RAMZI's) people are . . . overseas."

According to the 302, "Scarpa was then told to continue to attempt to gather information on the alleged conspiracy to blow up aircraft, but not to encourage the plot in any way."[26]

Earlier, White had noted that Murad had told Scarpa that such a bombing "is going to happen and they are checking to see if BOJINKA received the message."

This account was written thirty-four days before the downing of TWA Flight 800.

Eleven days later, Yousef told Scarpa that "he is not going to perform the airplane explosion for now because the trial is going well."[27] But three days after that, the mercurial Yousef told Scarpa that he thought the government "wants to sabotage his case." With his notorious ego, the bomb maker had felt sure that he could win by representing himself. Now, as the old adage went, it was beginning to look like he had "a fool for a lawyer." Still, whatever Yousef's shortcomings as an attorney, his connections to al Qaeda were as strong as ever: In addition to the phone link, he boasted to Scarpa that he had a way of getting written coded messages in and out of the MCC to his cohorts abroad, which included members of his family living in the Baluchistan region of Iran.

A Possible Al Qaeda Link to the Khobar Blast

On June 25, nineteen Americans were killed when a truck bomb tore through the Khobar Towers complex in the Eastern Province of Saudi Arabia. Though Iranian members of Hezbollah were later identified by the FBI as the perpetrators, Yousef told Scarpa that Bojinka—that is, bin Laden—was the "mastermind." In fact, Yousef boasted that he was originally sent "to check out the security measures and that a tanker truck [to deliver the bomb] was discussed at the time."[28] That last reference was from a Bureau 302 on July 2, 1996. In late June 2004, Dan Eggen of the *Washington Post* reported that the 9/11 Commission had determined that "Al Qaeda . . . may . . . have played a 'yet unknown role' in aiding Hezbollah militants in the 1996 bombing of the Khobar Towers."[29] It is unclear whether that new conclusion was influenced by Scarpa's revelations—evidence that had been brought to the Commission's attention as early as April 5, 2004, more than two months earlier.

By the time of the Khobar blast, apparently, Yousef was also preoccupied by thoughts of murder closer to home. Subject to mood swings, no doubt agitated by the progress of his case, Yousef had begun acting in what

Scarpa found a threatening manner. Yousef and Murad had learned the address of Scarpa's wife and children. Later Yousef joked with Scarpa that he would murder him by poisoning his food. Scarpa began to worry about the safety of his family.

One day, Yousef came back from court and renewed his pledge to kill ASUA Mike Garcia because he believed that the prosecutor had "smirked" at him. Now, although he still had the bomb maker's trust, Scarpa was beginning to feel like he was walking the edge of a razor blade. The way he saw it, he was risking his life for no guaranteed payoff. So he began to dial back on his relationship with Yousef.

By mid-July, AUSAs Mike Garcia and Dietrich Snell were ready to deliver a devastating blow to Yousef and his co-defendants Murad and Shah. They were about to enter evidence from Murad's confession to the PNP that he and the other Bojinka plotters had intended to plant Yousef's ingenious bomb triggers aboard up to eleven airliners leaving Asia for the United States.

The document, which described Yousef's desire to bring the jihad to Paris, would have amplified previous evidence entered by Snell and Garcia detailing the Casio-driven bomb that Yousef used on PAL 434 in 1994. This was the wet test that Yousef considered the Bojinka rehearsal. On the face of that confession alone, Murad, Yousef, and Shah might have been guaranteed a conviction.

Then, at: 8:18 P.M. on the night before the evidence was due to be presented, a 747-100—identical to the PAL airliner that Yousef had bombed—taxied for takeoff from JFK bound for Paris, carrying 212 passengers and a crew of eighteen.

Seconds after the plane cleared Long Island and leveled off at 13,000 feet, at 8:31:12[30] P.M., an unknown "event" occurred in the area of the center wing fuel tank near the 23–25 rows. Whatever that event was, it triggered the detonation of the fuel tank itself, and blew the plane out of the sky.

It was TWA Flight 800.

5

TWA 800: BOJINKA FULFILLED

Twenty-five minutes later, the beeper went off on the hip of James Kallstrom,* the assistant director in charge of the FBI's New York office.[1] As he sped downtown to 26 Federal Plaza, he contacted FBI director Louis Freeh, then called his own number two, Tom Pickard. Kallstrom instructed Pickard to get ahold of John O'Neill, the head of the FBI's counterterrorism unit in Washington.

O'Neill then called President Clinton's terrorism czar, Richard Clarke, and by 10:00 P.M. a meeting was convened in the White House Situation Room. The vast majority of commercial aviation crashes are tragic accidents, and there had never been a U.S. commercial aircraft downed by an act of terror within American borders, so this level of White House reaction was unusual. But it was clear on this night that virtually everyone in the official chain of command suspected terrorism. And when Pickard finally reached Neal Herman, the head of the FBI-NYPD's Joint Terrorist Task Force, the first words out of Herman's mouth were, "I'm thinking Yousef."[2]

At that point, of course, why wouldn't he? After all, for the last few months, a mob informant on the ninth floor tier of the MCC had been

* See picture, p. 267.

feeding the Bureau detailed intelligence predicting a bomb explosion aboard a plane. Yousef, the architect of the PAL bombing—a wet test for the Bojinka plot—had even fed Scarpa a blueprint of just such a device. Further, the Feds had actually facilitated calls between Yousef and his "people" in the Mideast, one of whom was his uncle Khalid Shaikh Mohammed, the last of the four Bojinka co-conspirators who was still at large, and one who knew how to build the Casio timed-nitroglycerine device.

The motive? Yousef had given it to Scarpa: he wanted a mistrial. And the next day, as Navy divers searched the still-burning wreckage and bodies began floating across Long Island sound, Yousef and co-counsel Roy Kulcsar effectively asked for one. In a bench sidebar with Judge Duffy before the start of trial, Kulcsar noted "the unfortunate confluence of circumstances" in the fiery crash, which bore such an eerie resemblance to a Bojinka event.[3]

But Duffy, the tough Bronx Irishman whose own life had reportedly been threatened by Yousef, was in no mood to cut him any slack. He promised to poll the unsequestered jurors to see if they had been unduly influenced by the nonstop TV coverage and banner headlines of the disaster. Then he went even further. Minutes later, when he'd brought in the jury, Duffy made a finding of fact that was in direct conflict with the very evidence the FBI had been gathering from Scarpa over the previous two months:

"Good morning, ladies and gentlemen," the judge began. "Last night near Moriches Inlet out in Long Island, an airplane blew up. TWA Flight 800. Now there is going to be . . . all kinds of speculation about what happened. I have no clue what happened, nor do you, nor do any of the people who have been speculating up to this point. All we know is that there was an explosion and the airplane went down. It's a tragedy, there is no two ways about it, but that had nothing to do with this case."[4]

The court transcript doesn't reveal whether Assistant U.S. Attorney Mike Garcia, whose life had been threatened by Yousef, exchanged looks with his co-counsel Dietrich Snell. We don't know whether Patrick Fitzgerald chief of the SDNY's Organized Crime-Terrorism unit, was in the courtroom. But any Fed who had access to Scarpa's warnings had to know that Judge Duffy was dead wrong: the crash of TWA 800 had *everything* to do with the Bojinka case.

The FBI would soon find evidence of high explosives in the passenger cabin on the right side of the aircraft. They would find damage consistent with an explosion by one of Yousef's Casio-nitro bomb triggers, set underneath a seat somewhere in rows 17–25 over the center wing fuel tank—the "event" that had detonated the fuel tank and decapitated the plane. In his wet test aboard PAL 434, Yousef had been one row short in his placement, and succeeded in killing only a single passenger. Now, with a shift of perhaps no more than a row or two, 230 passengers were dead. The crash of TWA 800 at a crucial moment in the trial was Bojinka fulfilled.

Unaware of the FBI's secret informant Greg Scarpa or the intelligence he was furnishing to the Bureau, the *New York Times* sided with Judge Duffy, declaring in a story on July 23 that "There is no evidence to link Mr. Yousef, who has been jailed here for nearly a year and a half, to the crash of the Trans World jetliner." But *Times* reporter Christopher S. Wren left an opening: "Law enforcement officials," he wrote, "are examining whether any of Mr. Yousef's sympathizers on the outside might have bombed the plane to protest his detention and trial."[5]

In her flattering biography of James Kallstrom and the TWA 800 investigation, *In the Blink of an Eye,* Associated Press reporter Pat Milton wrote that after the crash "agents confirmed no actual threats had been called into the FBI."

Yet this investigation has found that several Islamic groups took credit for the crash—including one seven hours before the plane exploded. At about that time, an Arabic daily in London received a fax from a radical Muslim group, the Islamic Movement for Change. This was the same Iranian-related group that took credit for the 1995 bombing of the U.S. military training mission to the Saudi National Guard in Riyadh, which left five Americans dead. The message warned that "The Mujahedeen will deliver the harshest reply to the threats of the foolish American President."[6]

According to Attorney General Janet Reno, after the crash two groups also claimed responsibility. One, identifying itself as a fundamentalist Islamic group, contacted a Tampa, Florida, TV station.[7] But an even more significant claim of responsibility was uncovered on July 31, 1997. When the FBI and NYPD raided the Brooklyn apartment of two Islamic radicals accused of attempting to bomb New York's subways, they recovered a note that not only took credit for the downing of TWA 800, but demanded the release of Ramzi Yousef.[8]

In yet another unexpected convergence in a story ripe with them, the judge presiding at the trial of the reputed subway bombers, Ghazi Ibrahim Abu Maizar and Lafi Khalil, was Reena Raggi—the same judge who years earlier had ordered Mohammed Ajaj's bomb manuals returned prior to the WTC bombing. (Several months later, in the fall of 1998, Raggi would preside over the case of Greg Scarpa Jr. himself.)

Perhaps the most telling evidence of Yousef's connection to the crash, however, came in an FBI 302 memo dated July 24, 1996. A week after the plane went down, FBI agents who interviewed Greg Scarpa wrote: "SCARPA advised that before the cell rotations MURAD stated that he feels that they may get a mistrial from the publicity surrounding the TWA explosion."[9]

Evidence of a Bomb on Board

Although a number of witnesses on Long Island reported seeing a white arc of light in the sky just before the explosion, the crash of TWA 800 had all of the earmarks of Bojinka. The cockpit voice recorder showed no hint of any mechanical cause for the crash.[10] There had been no mayday from the crew. After the event happened at 8:31:12, the plane continued climbing for 24 seconds before the entire front end blew away.[11]

Suspecting a crime, FBI ADIC James Kallstrom immediately dispatched hundreds of agents to the scene off East Moriches Point on Long Island. Almost from its arrival, though, the team from the National Transportation Safety Board began pushing a mechanical theory to explain the crash.

"We treated it as a potential criminal matter from the beginning," says Kallstrom (now retired) in an interview for this book. "I would have bet my meager government salary at the time that it was a crime."[12]

As a result, a pitched battle broke out between the two federal agencies. The tension became so pronounced that by July 21, U.S. Attorney Valerie Caproni, chief of the Eastern District's Criminal Division, informed NTSB investigator Bruce Magladry that "no interviews were to be conducted by the NTSB." Safety board investigators would be allowed to review FBI supplied documents on the witnesses, she ordered—but only "provided no notes were taken and no copies were made."[13]

It should be noted that AUSA Caproni was the same prosecutor who issued the memo assuring that there would be no "spy in the defense camp" during the Bojinka trial as the result of Greg Scarpa's intelligence. As it turns out, she had played a pivotal role in the series of Colombo War prosecutions that would later affect Scarpa Jr.'s fate. At this point in the summer of 1996, however, Caproni and James Kallstrom were both focused on the TWA 800 crash.

The theory that the 747 was downed by a missile from a nearby U.S. Navy exercise on the night of the crash was soon discounted, as large sections of the wreckage* were assembled at the former Grumman aircraft hangar in Calverton, Long Island—ironically, just minutes away from the rifle range where the FBI's SOG surveillance team had photographed members of Yousef's al Qaeda cell firing weapons seven years before.

The NTSB believed that the center wing fuel tank exploded after fuel vapors were somehow ignited. They pointed to an incident in the Philippines in which a Boeing 737 reportedly exploded on the airport tarmac, killing eight passengers, after a defective float valve tied to frayed wiring caused an explosion of vapors in the partially empty tank.

Based on the intelligence coming from Scarpa, though, Kallstrom, the head of the Bureau's New York office, wouldn't give up on the bomb theory. In fact, he was so impressed by the quality of Scarpa's notes from Yousef's kites that he personally telephoned Scarpa's attorney Larry Silverman just days after the crash.

He said "Don't let Scarpa stop now," remembers Silverman.[14] "It was at a time when Greg was starting to feel—rightfully so—that whatever he did would never get him a benefit and that while he may be helping others, it wasn't helping him at all. And he was putting his family and himself in significant jeopardy. But Kallstrom made it clear to me that it was his perception that the information that was coming from Yousef and the

* See picture, p. 270.

other people [through Scarpa] was critically important and that he didn't want to let it evaporate."

The idea of an assistant FBI director calling a defense lawyer in the middle of a crash investigation was highly unusual. But Kallstrom says he'd seen summaries of Scarpa's evidence from Yousef. "We had this information from Scarpa," he says, "so the likelihood was, without any mechanical reasons front and center, that it was a criminal act." Soon, as wreckage was retrieved from the bottom of Long Island Sound, Kallstrom's instincts appeared to be directly on target.

In July, traces of the high explosive PETN were found on the plane's right wing near the mid-section, close to the point of detonation.[15] Landing gear recovered at the site showed signs of having been blown out from the blast.[16] Two metal seats in the twenty-third row had fist-sized holes blown through them;[17] and days later, RDX was discovered near the PETN above the center wing fuel tank.[18]

Yousef had recommended the use of RDX in his May 19 kite to Scarpa, under the subhead "How To Smuggle Explosives Onto An Airplane." In fact, within days of the discovery of RDX in the TWA 800 wreckage, Steven Burmeister, head of the FBI's lab in Washington, testified at the Bojinka trial that investigators had found methenane, a key component of RDX, in Yousef's Manila bomb factory.[19]

But the clincher seemed to be the discovery of nitroglycerine in the passenger cabin, thirty feet away from the PETN.[20] A diluted nitro solution was the key component of the Bojinka wet test device that Yousef hidden in the life jacket pouch beneath seat 26K on PAL 434. By mid-August, the *New York Times* was even quoting FBI investigators as theorizing that a bomb might "have been hidden beneath a seat."[21]

In the early stage of the investigation, engineers at Boeing disagreed strenuously with the NTSB's fuel-tank theory, arguing that combustion in the tank would have been on the order of fifteen pounds per square inch and insufficient to blow the big jumbo jet apart.[22] But the FBI's chief metallurgist, William Tobin, sided with the Safety Board.

With less than half of the wreckage recovered, Tobin declared that neither a missile nor a bomb had brought down the plane.[23] "'No indicia of high explosive' was the way he put it," remembered Kenneth Maxwell, the

FBI's supervisory special agent who ran the Calverton site. "It was his position that there was no bomb and he was the ruling authority."[24]

Tobin threw his weight behind the NTSB investigators, who were speculating that the center-wing fuel tank explosion was the result of a fuel vapor ignition caused by a short in the wiring. The metallurgist found none of the trauma in the wreckage consistent with the kind of high explosive blast that had downed Pan Am Flight 103 over Lockerbie, Scotland. But that plane had been detonated by almost a pound of Semtex, an RDX-related Czech form of the explosive C-4. Hidden in a Toshiba boom box, that bomb had been powerful enough *on its own* to destroy the Pan Am 747.

Yousef's twisted genius was in using his Casio-timer bomb trigger as a kind of blasting cap to ignite the center fuel tank—so the resulting damage would never bear the earmarks of a Pan Am 103-like event. And Maxwell, who eventually oversaw the assembly of 96 percent of the plane's wreckage,[25] soon locked horns with Tobin, an FBI Headquarters authority with thirty-five years' experience.

In an interview for this book, Tobin said that he couldn't remember whether he had considered the forensic evidence from Yousef's bomb aboard PAL 434 when judging whether TWA 800 had sustained damage consistent with an explosive device. But Maxwell, who oversaw the Calverton investigation, suggested that he hadn't. "What Tobin was laboring to do was find indicia of high explosive damage," he said. "With no comparative database."[26]

In an earlier interview, Maxwell reiterated that Tobin had "made up his mind early on in the case that it had to be an accident because he didn't see any blatant metal evidence."

Still, by August 22, the *New York Times* seemed ready to conclude that TWA 800 was brought down by a criminal act.[27] Reporter Don Van Natta Jr. was about to file a story saying that "investigators have finally found scientific evidence that an explosive device was detonated inside the passenger cabin of Trans World Airlines Flight 800." Quoting "senior Federal officials," Van Natta confirmed that FBI chemists "had found traces of PETN, a chemical in plastic explosives, on a piece of wreckage retrieved from the jet's passenger cabin between rows 17 and 27." Those ten rows above the center wing fuel tank were described by Van Natta as "the epicenter of the blast."

Noting that "in loss of life, the downing of TWA Flight 800 would stand as the most serious crime in American history," Van Natta reminded readers that on July 29 Joseph Cantamessa Jr., a special agent in charge from the Bureau's New York office, had said that "one positive result" in forensic tests would be enough to declare the crash a criminal act.

There was even published speculation elsewhere that a bomb could have been placed aboard the fatal flight in the Athens airport, where it had been serviced before departing for JFK. The FBI would later announce that it was gearing up to send agents to the Greek airport, which had been reportedly cited in the past for "lax security."[28]

Van Natta seemed so impressed by the PETN discovery that he noted how "for weeks, criminal investigators have said they would need positive findings of explosive residue at the Washington lab before they could conclude what most of them have believed all along—that a bomb, not an unusual mechanical malfunction, destroyed the jet shortly after it left Kennedy International Airport."

Interviewed for this book, a confidential FBI source said that Kallstrom himself was ready to announce that the TWA 800 crash had formally become a criminal investigation. The retired ADIC himself denied this vigorously in an interview for this book, insisting that he kept his mind open throughout the probe.

But on August 22, the day before the *Times* story broke, Kallstrom was summoned to Washington, D.C., for a high-level meeting with FBI Director Louis Freeh. Also present was Deputy Attorney General Jamie Gorelick, who would go on to become one of the ten 9/11 Commissioners.[29] Until that meeting, the FBI's declaration of the TWA 800 crash as a crime seemed to be a foregone conclusion. But something was discussed among those top FBI and Justice Department officials that caused the Bureau to make a sudden 180-degree turn.

By the time he returned to New York that day, Kallstrom was on the phone to Van Natta and the *Times* editors trying to get them to kill the bomb story. According to the AP's Pat Milton, the FBI ADIC now felt that "to publish [the PETN] story would be a rush to judgment."[30]

The next day, when the *Times* ran Van Natta's piece under the headline "Prime Evidence Found That Device Exploded in Cabin of Flight 800," Kallstrom read a statement confirming the PETN discovery. But he cautioned that it was premature to link the explosives to the crash. Seeming to side with metallurgist Tobin and the NTSB, Kallstrom was now demanding "additional proof, such as explosive shock waves in salvaged metal."[31]

"For whatever reasons," says the FBI source, "the Bureau and Justice were gearing up to spin this thing away from a bombing plot to some kind of mechanical event."[32]

A day later, on August 24, Kallstrom went out of his way to disassociate the PETN from the crash. He was quoted in the *Times* as saying that "it's not inconceivable that this chemical would be available through some other means other than through an explosive device, and left on the airplane." But a senior law enforcement official also interviewed was said to have "laughed out loud" at the suggestion. According to *Times* reporter Dan Barry, "investigators did not take such a scenario seriously."[33]

Soon, however, word went out asking FAA field offices to seek an alternate explanation for how high explosive residue might have appeared on the Flight 800 wreckage *absent* an explosive device. CNN reported that the FBI was now claiming that the PETN/RDX residue "could have been brought on the plane by a passenger and was not part of a bomb."[34]

By August 27, FBI investigators were speculating that perhaps "the chemical traces might have been brought aboard by soldiers who flew the plane when it was pressed into service during preparations for the Persian Gulf War," five years earlier.[35] But that theory was scotched when CNN learned that the plane has been refitted from nose to tail after its last use by the military.[36] In late August, Kallstrom asked George Andrew, the special agent who worked directly with Tom Pickard, to go back and ask the FAA if any live explosives had ever been placed aboard the fatal aircraft to train bomb-sniffing dogs.[37]

Then, on September 19, the Bureau got a hit. It seemed that five weeks before the crash, on June 10, an officer from a K-9 unit of the Lambert-St. Louis International Airport police had conducted an explosive training exercise with a bomb-sniffing dog on a wide-body jet at the end of the C Concourse. The Bureau told the *New York Times* that the use of high explosives in the test "could explain the traces of explosives found in the wreckage" of Flight 800.[38]

According to the *Times,* the "test packages" used in the K-9 exercise "contained the same explosives found by investigators after the crash." Despite the fact that the piece was written only a day after the St. Louis discovery, *Times* reporter Matthew Purdy quoted "law enforcement officials" as declaring that "the testing of the dog on the plane would make it nearly impossible for investigators to construct a circumstantial case that the plane was destroyed by an explosive device."

Even though TWA officials refused to confirm that the "test" plane was the aircraft that became TWA Flight 800, Purdy reported *as fact* that "investigators were able to identify that the plane involved in the test was the same one that crashed off the coast of Long Island."

The next day, Sunday, September 22, Don Van Natta, who had written a month earlier that the FBI was a hair's breath away from calling the TWA crash a crime, was now declaring in the *Times* that "the training exercise casts serious doubt on the best evidence of an explosive device yet found."[39] He described the last-minute revelation of the K-9 test as "a stunning setback," and noted that "When he found out about it on Friday, one investigator said that the news hit him like 'a punch in the gut.' " Another official was quoted as saying, "We're back to square one."

Within weeks, suspicion became fact. The bomb theory was dead. Kallstrom himself went on to suggest that the residue from the dog test was left on the aircraft due to the officer's negligence: "An incredible amount of this chemical leaking out of these packages fell into that spot," said the ADIC in Congressional testimony, referring to the area of the wreckage were the PETN and RDX had been found.[40]

By 1997, John O'Neill, the FBI's chief counterterrorism official, who at first had suspected a Yousef connection, had been battered into accepting the mechanical theory. Even President Clinton's terrorism czar, Richard Clarke—perhaps the most knowledgeable man in Washington when it came to the bin Laden threat—accepted the fuel tank ignition scenario.

"In the days that followed," wrote Clarke in his memoir *Against All Enemies,* "no intelligence surfaced that helped advance [a criminal theory in] the investigation."[41]

The president himself was convinced. In his biography, *My Life,* Bill Clinton wrote: "On July 17, TWA Flight 800 exploded off Long Island, killing some 230 people. At the time everyone assumed—wrongly, as it turned out—that this was a terrorist act."[42]

In another note of irony, John O'Neill, the FBI's chief counterterrorism official, came to believe that the TWA 800 investigation was draining valuable Bureau resources from the war on terror. By the time he transferred from FBI Headquarters to run the National Security Division in New York, O'Neill started talking to ADIC Kallstrom about "an exit strategy."

Neal Herman, the JTTF chief who pegged Yousef on day one of the crash, was later quoted as asking "My God, what the hell is an exit strategy? We get out of cases when they're over."[43]

"O'Neill's discussions with Kallstrom and with Washington and the NTSB were putting on a lot of pressure," said Supervisory Special Agent Maxwell. So by mid-November, sixteen months after the crash, Maxwell got a call from Kallstrom.

"'Shut it down,'" Maxwell recalls him saying. "It was a very direct order. 'Shut it down.'"

James Kallstrom, the FBI's assistant director in New York, a man who had found Greg 's intelligence from Yousef so "critically important" that he called Scarpa's lawyer to encourage him, was now not only *accepting* the NTSB's mechanical theory for the crash, he was *endorsing* it. In a letter sent to victims' families on November 15, the Bureau went so far as to state that it had "found absolutely no evidence" that the plane was downed by a criminal act.[44]

Despite the fact that in the millions of flight hours the 747 had been in service, no center fuel tank had ever exploded in mid-air; despite the fact that the only comparable event took place on a 737, which caught fire on the tarmac in a Philippines airport after evidence of a short in a defective float valve; despite the fact that there wasn't a scintilla of proof that a frayed wire or a voltage overload had caused a short onboard Flight 800, the NTSB prevailed. The official cause of the crash that killed 230 people was determined to be "mechanical failure."

Only a handful of people inside the FBI and the Justice Department knew the truth: that a mob informant had uncovered extraordinary evidence that al Qaeda cell was responsible for the crash. That Greg Scarpa Jr., a man with a limited education, had furnished the FBI with detailed bomb formulas identical to the device that Yousef had exploded on an identical 747-100; that Yousef himself had boasted that he would "explode a bomb" aboard a U.S. airliner to get a mistrial; and that he had even offered to connect the Feds with his cohorts overseas—who, history would soon show, were already plotting the 9/11 attacks.

But there was still some insult to be added to the injury. When it came time for the Feds to reward Scarpa for risking his life and that of his family by informing on the world's most dangerous terrorist, the Justice Department reneged.

Refusing to grant Scarpa Jr's 's 5K1.1 downward departure request, they called his year-long intelligence initiative with Yousef "a hoax" and "a scam." Not only was Scarpa not *rewarded* for his work, the Feds decided to throw the book at him.

In light of what we now know from the Yousef-Scarpa kites and the FBI 302s that authenticate them, a series of questions remain:

Why did a veteran agent like James Kallstrom, now terrorism advisor to New York governor George Pataki, change his mind? What persuaded him to defy the weight of the explosive's evidence and go with a mechanical crash theory, one that to this day has never been scientifically proven? And why did Kallstrom allow the St. Louis K-9 test to carry so much weight?

That question becomes critical in light of the new evidence uncovered in this investigation, which invalidates the FBI's explanation of how explosives got on the plane.

This additional new evidence now demonstrates that the test done in St. Louis five weeks before the fatal crash was *not* conducted on the 747 that became TWA 800, but rather on a sister 747 that looked almost identical; that the only explanation by the FBI and the NTSB for the presence of high explosive residue on the flight wreckage was *wrong;* and that, in fact, the K-9 officer whom James Kallstrom accused of allowing his test aids to "leak," did his exercise on a different plane.

6

SHATTERING THE K-9 THEORY

In a September 1997 letter to former congressman James Traficante (D-OH), who was questioning the FBI's probe of the TWA 800 crash, FBI ADIC James Kallstrom wrote the following:[1]

> On September 20, 1996, a patrolman for the St. Louis Airport Police Department (SLAPD), who is assigned to the canine unit was interviewed by FBI Agents in St. Louis. The patrolman advised that his responsibilities included maintaining the training for his explosives sniffing dog on a daily basis so that he could meet FAA requirements for training and certification. According to the patrolman, it is normal procedure to conduct training for the dogs on virtually a daily basis on available aircraft.
>
> On the morning of June 10, 1996, while working the day shift at the St. Louis International Airport, the patrolman placed a call to the manager on duty at TWA Line Service to determine if they had an aircraft available on which the patrolman could conduct some training for his bomb sniffing dog. The manager on duty, whose name the patrolman could not recall, told him that a "wide body" was available at, gate 50 at the St. Louis Airport and that the patrolman could use this aircraft to conduct his training. The patrolman recalled that he was particularly enthused because it is rare that "wide body" aircraft become available for such training at St. Louis.
>
> The patrolman retrieved four types of explosives from the SLAPD explosives bunker for use in the training. The explosives retrieved were water gel, C-4, det

cord and ammonia dynamite. He also used smokeless powder, which was stored in the trunk of his patrol car, in the training.

The officer noted that the explosives bunker contains a variety of military and commercial type explosives for use in training and opined that the bunker would very likely contain residue of these explosives. After retrieving the explosives, the patrolman proceeded in his patrol car to Gate 50 where he found a 747 parked.

The patrolman parked his vehicle at the base of the stairway at the outside of the jet way and entered the aircraft. The patrolman determined that the electric power was on and that no one else was present on the plane. He returned to his patrol car and to bring the explosives on board the aircraft, which he believes he accomplished in two trips. The explosives were initially placed on the counter in the galley just inside the main entry door to the aircraft. The patrolman then proceeded to place the explosives around the aircraft interior for the training/certification exercise.

The patrolman proceeded to place the explosives about the aircraft as follows:

1. The smokeless powder was on its side with the cap unscrewed inside the center armrest of row 2, seat 2 of the first class section.
2. The water gel was placed on the floor inside a tall, narrow closet/storage bin at the rear of the upper level first class section.
3. A 1.4 pound block of C-4, covered with a thin covering of clear cellophane type material, which the patrolman described as being in poor condition and allowing some of the explosive to be exposed, was placed in the pouch on the back of the backrest of row 10, seat 9.
4. The det cord, which was described by the patrolman as at thirty foot piece in extremely poor condition with cracks every few inches, was brought in its container to row 20 of the main cabin. The patrolman said that he believes he went to the side of the cabin opposite from the side where he placed the C-4 since it was his practice to place the explosives in a zig-zag pattern within the aircraft. The patrolman placed the container in which the cord was stored on the floor in the aisle, removed the cord and placed it in an overhead compartment in row 20. The patrolman noted that the can containing the det cord contained quite a bit of powder from the det cord and said if one were to wave it in the air it would create a visible cloud of powder.
5. One stick of ammonia dynamite was partially concealed in a groove in the flooring near an emergency door labeled "PRE" on the same side of the aircraft as he placed the C-4. The patrolman believes the door was located over the wing.

The patrolman advised that he began the placement of the explosives at 10:45 A.M. and is required by FAA regulation to wait 30 minutes from the first placement before commencing the training exercise with the dog.

At 11:45 A.M., the patrolman began the exercise by bringing the dog into the aircraft and working him through the three areas of the aircraft where the explosives were placed. The exercise lasted fifteen minutes and the dog located all the explosives.

After returning the dog to his patrol car, the patrolman proceeded to remove the explosives from the aircraft in the same order in which he placed them, using the galley as the center of his movements.

The patrolman stated that he did not enter any areas of the aircraft other than those described and specifically stated that he did not enter any cargo areas. He also stated that he was the only person involved in the exercise. The patrolman provided the FBI with the can of smokeless powder used in the exercise and advised that all the other explosives had been replaced by either the FAA or exchanged locally for fresh material since the time the exercise was conducted.

The FAA in St. Louis provided the FBI with a copy of a TWA document listing gate assignments for June 10, 1996. This document, a copy of which is attached, shows that a 747 bearing tail number 17119, which is the tail number for the 747 that was Flight 800, was parked at gate 50 from shortly before 700 hours (7 A.M.) until approximately 1230 hours (12:30 P.M.) on that date.

A copy of the same TWA gate assignment document provided to Traficante, marked to indicate the presence of two 747-100s on the C concourse on the morning of June 10, 1996, and the respective departure times of the two aircraft early that afternoon, can be found on page 312.

The first dot and line shows that 17119, the 747-100 that became TWA Flight 800, was parked at gate 50 until 12:35 P.M. An identical 747-100, designated 17116, was on the other side of the C Concourse, at Gate 52; it departed at 1:45 P.M. (See illustration of the concourse on p. 313.)

The officer conducting the K-9 test that day was Herman Burnett, a veteran of the SLAPD who was using the test to qualify a dog named Carlo. It was not Burnett's practice to record the "N" or tail number of the plane when he conducted explosive detection exercises. Nor was an ID required by the FAA. The only prerequisite was that the aircraft be empty and available for use for a period of up to four hours prior to departure.

By the FBI's own admission, Burnett concluded the test by noon, and it took approximately fifteen minutes to return the dog and all the explosive test aids to his vehicle, which was parked below the gates at the end of the C Concourse.

In 1996, TWA had a standing policy that its flight crews must arrive on an aircraft at least ninety minutes before departure.[2] In an interview for this book, Burnett told me that he did the test on an empty plane—but that as he finished, flight attendants were just beginning to arrive.[3]

His latest recollection conflicts somewhat with the FBI's account of the interview ADIC Kallstrom reported to Congressman Traficante. In an email to me on May 5, Burnett said, "When I asked TWA for an aircraft to train on, they gave me the wide body that was at the gate and advised me that I had to be off by 12:00 noon, which I was. I told the FBI this during the interview.

"In the [FBI] report you sent me, they have me waiting from the time I placed the first aid at 10:45 A.M. until 11:45 A.M., when I started working the dog. We only allow thirty minutes set time. So around 11:15, I started working the aircraft with the dog. I knew that I didn't have a lot of time to cover the whole plane and work the dog as usual, so I kinda took him from aid to aid." In a follow-up phone interview, he was even more insistent: "When I asked for the plane," said Burnett, "TWA said, 'you can use it as long as you're off by 12:00 P.M.' I said, 'No problem. It's just me and the dog.' I was off by noon. I'm certain."[4]

The second line on the Gate Assignment document shows that 17119, the fatal aircraft, pulled back from the gate at 12:35 that day, bound for Honolulu.

By Burnett's best account, his departure by noon would have left little more than than half an hour for the jumbo jet to be fully crewed, fully catered, and loaded with passengers before pulling back from the gate. Given TWA's policy of requiring the crew to board an hour and a half before departure, the math doesn't add up. Whether you consider the FBI's official report from ADIC Kallstrom or Officer Burnett's most recent recollection, there was simply not enough time for him to have conducted his explosives test on 17119, the aircraft that became TWA 800.

However, on the other side of the C Concourse at Gate 52 that day, the 747 with the tail number 17116 didn't leave the gate until an hour later, at 1:45 P.M. That aircraft stood empty for several hours after its arrival at 7:00 that morning. It was almost identical to the TWA Flight 800 aircraft:

Indeed, in October, 1971 it had rolled off Boeing's assembly line at Everett, Washington, two planes before it.[5]

Meeting Officer Burnett Face to Face

On May 4, 2004, I flew to Lambert-St. Louis International Airport and conducted an interview with Officer Herman Burnett. It was followed by several e-mails between us and conversations by phone from his home. In the course of those interviews, Burnett said that he told local FBI agents back in September 1996 that he couldn't be sure which 747 he did the test on—but they *insisted* that it must have been the aircraft that became TWA 800. Furthermore, he stated firmly that in conducting the K-9 test on June 10, 1996, he never spilled or allowed any of the explosives to leak, as Kallstrom had suggested.

"I was pushed in a lot of directions back then," he said.

Burnett, a dedicated officer with twenty-six years on the job, told me that he feels he's been used as a "scapegoat" and now, as he nears retirement, he wants to "set the record straight."

"The truth is that I honestly can't say [with absolute certainty] which plane we did the test on that day," he said in our airport interview. "But my notes, and my memory at the time the FBI talked to me, told me that it had to be the other plane. The one at Gate 52. I just wouldn't have had time to do the test and finish by the time that [the TWA] 800 plane took off. The feeling is, that it was a whole different plane altogether."[6]

John Nance, a decorated former Air Force pilot, lawyer, and author who is considered an expert in aviation safety, says that the FBI's K-9 test explanation for the presence of explosives aboard TWA 800 has always "puzzled" him.[7]

"The FBI found evidence of PETN and then they came up with what sounded like a cockamamie story that came about because they had a bomb dog test on board," says Nance. "That explanation has always left me cold. First of all, if that was the case, what was the residue doing on the wreckage? It should have been in an enclosed package. Secondly, how did it survive the accident, since the wreckage was under water for weeks?"

Keep in mind that, for the FBI's cover story to be plausible, Officer Burnett would have had to have been *negligent* in the conduct of the test. But he insisted that he wasn't; that he didn't "spill" any of the explosives as he put the dog through its paces.

"I was very careful with the test aids," Burnett told me during our meeting. Later by phone he reiterated that "nothing spilled."

But the legend of Burnett's alleged negligence seemed to grow with each official rendering of the test. In a February 1997 letter from NTSB chairman James Hall to Barry Valentine, then the acting FAA administrator, Hall wrote, "the dog handler had spilled trace amounts of explosives while placing training aids on Board the aircraft."[8]

"Based on interviews with the dog handler," Hall said, "the Safety Board determined that he had conducted the training exercise without taking adequate time and precautions when handling the explosives training aids."

Jack Cashill and James Sanders are investigative reporters who have always espoused the missile theory of the TWA 800 crash. But in their book *First Strike: TWA Flight 800 and the Attack on America,* their reporting on the unexplained presence of RDX, PETN, and nitro in the wreckage underscores key points. They noted that by the time the Associated Press's Pat Milton wrote her history of the crash, *In the Blink of an Eye,* her "retelling was even more defamatory" when it came to Officer Burnett.[9]

According to Milton, Burnett told the FBI, "Yeah, I could have spilled more than a little. The packages were old and cracked and we hadn't used them in awhile so more than usual might have come out."

But Burnett insists that the "explosive aids" did not leak. "I never lost any," he told Cashill and Sanders "I never spilled any. There was never any powder lying loose. I just hate that they twisted my words. I know what I did and how I did it."

An in an email to me on May 7, 2004, Burnett confirmed his story to me once more. But today he is worried—concerned that if too much light is focused on him, it could affect his retirement and his pension. "I just don't want all this to come back and hurt me now after all these years."

Why would a police officer who had conducted a dog-training exercise in the course of his duties in 1996 be worried about reprisals?

Officer Burnett has reason to worry—his word is all that stands between the FBI's theory of whether the crash of TWA 800 was an accident or a mass murder. If he didn't conduct the K-9 test on the fatal plane, then the FBI is left with no plausible explanation for the presence of high explosives—two of which were associated directly with Ramzi Yousef.

"The Bureau not only needs Burnett to have done that test," one FBI source told me, "They need him to have conducted it negligently."

And over the years, the official account of Burnett's purported carelessness has grown more and more embellished with each telling of the story.

In testimony before a congressional hearing in 1997, Kallstrom reported that "an incredible amount of this chemical leaking out of [Burnett's] packages fell into that spot" where the explosive residue was found.[10]

In his "Report to the American People on the Work of the FBI 1993–1998," former director Louis Freeh* underscored the importance of the K-9 test in the official explanation of the TWA Flight 800 crash.

> In 1997, the FBI announced that its sixteen-month investigation of the crash of TWA Flight 800 into the ocean off New York City had produced no evidence to indicate that a criminal act was the cause of the tragedy that killed all 230 persons aboard the airliner. . .
>
> At one point, traces of high explosive chemicals were found on the wreckage. The FBI then learned that in 1996 St. Louis Airport Police had conducted canine explosives training aboard the aircraft. Explosives put on board the plane at that time were consistent with the traces found in the postcrash tests. This was a major matter, and all of our tests and examinations show that the FBI got it right—the traces were from the earlier canine testing and not from a bomb aboard the aircraft.[11]

But the evidence now shows that the FBI didn't "get it right"—that it pressured a St. Louis Airport policeman into a story that can't be supported by either the FBI's own timeline or TWA's official gate assignment documents. In its effort to make the K-9 test fit the explosive residue, the Bureau apparently thought nothing of maligning the reputation of a veteran officer. Even FBI investigators who worked the probe have their doubts about what has now become "the single bullet theory" of the TWA 800 crash.

*See picture, p. 267.

The fact is that the test samples of K-9 exercise explosives placed by Herman Burnett in the zig-zag pattern throughout the passenger cabin didn't coincide with the discovery of explosive residue on the Flight 800 wreckage. On page 313, an illustration prepared by Tom Shoemaker,[12] a TWA 800 researcher, compares the FBI's description of Burnett's placement of the explosives with the PETN and RDX confirmed by the Bureau. "They're in completely different sections of the aircraft," says Shoemaker.

In fact, the FBI found the high explosives between the seventeenth and twenty-fifth rows, just over the center wing fuel tank on the right side of the aircraft, in a pattern consistent with the placement of one of Yousef's Casio-nitro bomb triggers. More important, none of officer Burnett's "aids" would explain the presence of *nitroglycerine* in the wreckage—the key explosive in Yousef's PAL 434 wet test device.

To make matters worse, no one in the FBI or the NTSB has ever been able to explain the presence of RDX on a curtain in the aft cargo hold, an area of the 747 that Burnett never visited. That fact troubles Ken Maxwell to this day.

"The RDX that was found on the aft cargo bay curtain is still unexplainable,"[13] said Maxwell. "It's still a mystery to me," he said, "what the ignition source was for the center wing tank. If you read the NTSB report, no one has ever identified the ignition source."

"The center fuel tank," says aviation safety expert John Nance, was "the engine of the destruction of the keel beam and the forward box—the force that unzipped the airplane. . . . There has never been an adequate explanation for what caused that tank to blow up."

The explanation, of course, should have been clear to anyone familiar with the evidence the FBI had in its possession—evidence that points unmistakably to an explosion that was deliberate, not accidental. Consider the aggregate of the facts:

- Ramzi Yousef had set a wet-test bomb trigger aboard PAL 434 in an effort to explode the center wing fuel tank and down a 747-100.

- Weeks before the TWA 800 crash, Yousef had given Scarpa a kite featuring a schematic of the bomb trigger he had used in the PAL 434 blast.

- Yousef had boasted to Greg Scarpa Jr. that his "people" would explode a bomb aboard an aircraft to seek a mistrial in the ongoing Bojinka trial in New York.

- The FBI had provided Yousef with the Roma Corp patch-through phone link, which gave him the ability to contact his uncle Khalid Shaikh Mohammed, the fourth Bojinka co-conspirator, who knew how to build the same kind of Casio-nitro device.

- The chemicals Yousef suggested to Scarpa were identical in nature to the chemicals found in the TWA 800 wreckage.

- The area of explosive residue coincided with Yousef's positioning of the PAL 434 bomb trigger.

- The NTSB's conclusion that an explosion in the center wing fuel tank was the ultimate cause of the TWA 800 crash.

- The new evidence on the K-9 dog test, which undercuts the FBI's official explanation for the presence of high explosive residue in the TWA 800 wreckage.

Taken together, the evidence strongly suggests that the first instincts of ADIC Kallstrom and JTTF chief Neal Herman were correct: the downing of TWA 800 was an act of al Qaeda terror—one that sapped the energy of hundreds of FBI agents and technicians for sixteen months, until their efforts were aborted in favor of a far less persuasive—but politically expedient—conclusion. This in turn caused John O'Neill, the FBI's top bin Laden specialist to believe that they were being diverted from the war on terror—when, in fact, they were right in the middle of it.

In light of these findings, the overpowering question is *Why?*

Why would a seasoned criminal investigator like James Kallstrom defy his own instincts and side with the NTSB, when they have never produced any definitive proof of a mechanical cause for the fuel tank explosion?

Why would he change course from near-certainty that a crime had been committed following the discovery of PETN, RDX, and nitro in the wreckage, to a hapless search for a cover story that turned an honest cop—

who believes he did the test on a different aircraft—into a negligent handler of high explosives?

Why would Kallstrom—who was floored by the quality of Scarpa's intelligence from Yousef—prematurely end an investigation into the murder of 230 people and acquiesce to a Justice Department conclusion that ultimately called it a "scam" and a "hoax?"

In the summer of 1996, at that critical point in the war on terror, why would the FBI turn down a chance to meet with active members of Ramzi Yousef's al Qaeda cell in New York and abroad?

Why not pay the $2,500 proposed by Yousef, allow FBI agents posing as mobsters to link with Yousef's cell members, and track them all the way to Khalid Shaikh Mohammed as he executed the hijack-airliners-suicide plot that culminated on September 11, 2001?

In short, why would the FBI's assistant director in charge of the New York office order the TWA 800 investigation to be terminated? What could James Kallstrom have possibly learned in that August 22, 1996, Washington meeting with FBI director Louis Freeh, attorney general Janet Reno, and deputy attorney general Jamie Gorelick (now a 9/11 Commissioner) that ultimately led him to call the man who was running the TWA 800 probe at Calverton and tell him to "shut it down"?

The evidence now suggests that the answers to those questions lay in an ends-justify-the-means decision made at the highest levels of the Justice Department and the FBI—a decision that concluded, in essence, that if Gregory Scarpa Jr. was found to be credible, a number of career-making criminal convictions might fall as a result. In the words of one Federal judge, up to nine cases might "unravel"[14] and James Kallstrom's New York FBI office would be rocked by the worst corrupt-agent scandal since Robert Hanssen was found to be selling secrets to the Russians.

The evidence now points to a cover-up that began in 1996, during the Clinton presidency, but remains intact to this day on the watch of George W. Bush—one that the 9/11 Commission itself was alerted to in the course of its investigation, but chose to ignore.

7

"THE ULTIMATE PERVERSION"

To fully appreciate why the Justice Department had abandoned Gregory Scarpa Jr., arguably one of the most important informants to date in the war on terror, we must first revisit the violent history of his father, who died of complications from HIV at the Federal Medication Center in Rochester, Minnesota, on June 8, 1994.[1]

The legacy Scarpa left behind would prove to be a bonanza for federal prosecutors. Before it was over, the so-called war started by the elder Scarpa had resulted in seventy-five separate prosecutions by the U.S. Attorney's office for the Eastern District of New York, with various Colombo family members indicted on charges ranging from racketeering and loan-sharking to drug dealing and murder.[2] Among those convicted were Pasquale Amato, a Colombo capo, and Victor "Little Vic" Orena, the acting boss and leader of the rival faction in the two-year war. By the time the smoke cleared, twelve people were dead.

At the time when Orena and Amato were sentenced to life, their attorneys had no idea that Scarpa had not only instigated the war, but had killed a third of the victims himself—and had done so with the apparent support and encouragement of R. Lindley DeVecchio, the FBI's top New York agent on organized crime.

Attorney Alan Futerfas and his associate Ellen Resnick were among the first to catch a whiff of this alleged corrupt relationship between the hit man and his "control" agent.

"Scarpa senior really wanted to become the head of the Colombo family," says Futerfas,[3] "So he used this opportunity to create a war to kill off all of his rivals. In doing so, his interests dovetailed perfectly with DeVecchio's, who saw this as a way to advance in the Bureau."

During the fourteen years he ran Scarpa as a Top Echelon informant, the flamboyant 6′5″ DeVecchio rose to the rank of supervisory special agent.[4] In defiance of FBI rules, which require two agents to run an informant (to help prevent just such corruption) "Del," as Scarpa called him, ran the Colombo captain by himself. The two men met two or three times a week, with DeVecchio often visiting Scarpa at home. At length they grew so close that they even vacationed together.[5] When Cabbage Patch Kids dolls were scarcer than bags of heroin, Scarpa reportedly furnished one to DeVecchio for his daughter.[6] Plying his control agent with fine wine and pasta,[7] he made sure that Greg Jr. had the champagne on ice when he paid for call girls to entertain the supervisory special agent at a Staten Island hotel.[8]

Bill Moushey, a reporter for the *Pittsburgh Post-Gazette,* conducted a two-year investigation of allegedly corrupt FBI agents that focused in part on DeVecchio. One of his pieces, published in 1998, carried the subhead "Federal agents sometimes fall prey to the lurid lifestyles of their informants."[9]

"Inevitably DeVecchio boasted that he knew how to 'talk, to act, as a true killer does,'"[10] wrote Frederic Dannen in a 1996 *New Yorker* profile, "The G-Man and the Hit Man." Various reports confirm that DeVecchio presented himself with the same bigger-than-life swagger as Scarpa, favoring Rolex watches, European silk suits with pocket squares, monogrammed shirts, and diamond cufflinks. "The word was that he married into money," said a former NYPD officer who worked on DeVecchio's squad. "None of us had a clue at the time just how close to Scarpa senior he was."[11]

After Scarpa's first control agent, Anthony Villano, closed down Greg Sr. as a source in 1975, he wrote a thinly veiled memoir of his years with the hit man called *Brick Agent,* underscoring the ends/means devil's bargain every FBI agent makes when working a mob informant. "I had to reassure myself," wrote Villano, "that our relationship was not the ultimate perversion of the whole law enforcement idea. In my mind, what we did was justified on the grounds of the greatest good."[12]

In 1980, after DeVecchio discovered Scarpa's name in the "closed" informant files of the New York office, he reopened him as an informant; soon, as Greg Jr. remembered, his father's relationship with DeVecchio was "better than with Agent Villano."[13] "Because of my father's close relationship with Agent DeVecchio," he wrote, "my father continued to receive immunity from prosecution and lenient treatment from the Government."

"Scarpa senior, there's little doubt, was a psychopathic killer," says defense attorney Alan Futerfas. "But he was given free rein by Lin DeVecchio. In the years they were together, he committed multiple murders, but Scarpa was never pulled over. He did less than a month in jail in fourteen years. I think DeVecchio knew exactly what Scarpa was doing. But he was writing [FBI] 209s [confidential informant reports] that would cover himself."[14]

Still, as early as 1987 the Feds were growing concerned that DeVecchio was sleeping with the enemy. By then, Greg Scarpa Jr. was involved in a violent Staten Island dope and extortion racket. Assistant U.S. Attorney Valerie Caproni* was on the verge of taking the Scarpa operation down, with the help of the DEA, when Greg Jr. showed his crew a list indicating they were about to be arrested.[15]

"Because of this inside 'tip' from FBI Agent DeVecchio," Greg Scarpa Jr. recalled, "I became a fugitive from justice for approximately eight months."

Caproni, then a young federal prosecutor, was clearly angered by the disclosure. "This was going to be a big expensive case to try and our top defendant was in the wind," she said.[16] "A Southerner with a reputation for toughness," as the *New Yorker* profile noted, Caproni—who went on to become chief of the EDNY's criminal division—later suspected DeVecchio as

* See picture, p. 267.

the source of the leak. In a sworn affidavit she wrote, "There is some reason to believe . . . that DeVecchio told Scarpa that the DEA planned to arrest Gregory Scarpa Jr."[17]

Caproni also suspected DeVecchio of leaking the name of Cosmo Catanzano, who Greg Jr.'s crew learned from a source was "going to tat." Greg Sr. ordered Catanzano killed and buried, and a grave was even dug for him on Staten Island, but he survived after being arrested.[18] In Dannen's *New Yorker* piece, DeVecchio vehemently denied both alleged leaks.

Yet the younger Scarpa confirmed them in 2003. In a sworn affidavit supporting defense efforts to overturn Columbo war convictions, he charged that DeVecchio passed on "personal information such as addresses and phone numbers to help my father locate people who owed him money for his illegal loan sharking activities and the identities of informants . . . who might be a threat."

Leaking the names of potential mob "rats" was one thing. But Scarpa Jr. went even further in his affidavit, charging that the supervisory special agent was also a co-conspirator to murder.

"With Agent DeVecchio's knowledge and consent," he wrote, "my father murdered or attempted to murder the individuals who were dangerous to him. My father would then report the successful killings back to the FBI, attributing to others responsibility for his murders. Committing murders and blaming others, always with the knowledge of Agent DeVecchio, became the 'M.O.' adopted by my father."

Not all of this, said Greg Jr., was done at arm's length. DeVecchio, he reported, was actually present when some of the felonies were committed.

"On several occasions," he wrote, "Agent DeVecchio personally monitored crimes committed by my father's crew. If law enforcement became aware of the crime while it was in progress, Agent DeVecchio was to prevent arrests by claiming that the participants included FBI informants. Agent DeVecchio provided this service for a least four robberies [of two banks and two furriers], in which I personally participated."

By 1992, Christopher Favo, the agent directly under DeVecchio in squad C-10 in the New York office, became concerned that key intelligence on the location of witnesses in the Columbo war was finding its way to Scarpa Sr.[19]

At one point Favo had given DeVecchio a partial address for the hideout of Vic Orena, Greg Sr.'s chief rival. He'd also furnished a mistaken address for one of Vic's soldiers.[20] Later, a cooperating member of Scarpa's crew admitted that he'd tried to kill both men by hunting them at the faulty addresses, which Greg Sr. had supplied to him from his "law enforcement" source.[21]

By March 2, 1992, the Bureau ordered DeVecchio to close Scarpa and terminate contact with him. Special Agent Favo later alleged that DeVecchio lied to FBI brass about his knowledge of the elder Scarpa's involvement in three separate homicides.[22] But DeVecchio wrote to FBI Headquarters asking for permission to reopen Scarpa Sr.

Incredibly, despite Scarpa's homicidal reputation, DeVecchio was given the okay. He would testify later that FBI supervisors "right up to the top" were aware of his relationship with the hit man, and that the SAC in charge of the New York office knew that Scarpa Sr. had committed "multiple murders," but permitted him to stay "open."[23]

"You have to understand what was going on," said an FBI source close to the investigation. "With Lin DeVecchio's help, Scarpa Sr. was wiping out by murder or arrest, the members of one of New York's last great crime families. It was a benefit for the Bureau. It just happened to wreak havoc with the constitution and the rule of law. But nobody at [FBI] headquarters was losing any sleep. And why should they? Until bin Laden came along, the Mafia was at the top of the Bureau's target list. In those days, pre-9/11, breaking the back of organized crime was good for a lot of points on Capitol Hill."

On May 22, 1992, Scarpa Sr. killed rival soldier Larry Lampesi with a shotgun; another Orena loyalist was wounded in the attack. When Special Agent Chris Favo reported both attacks to DeVecchio, he laughed, slapped his desk with his open hand, and exclaimed "We're going to *win* this thing."[24] Favo testified later that DeVecchio "seemed to be a cheerleader for the Persico faction" in the war and that "a line had been blurred. . . . He was compromised. He had lost track of who he was."[25]

Meanwhile, by late spring of 1992, as the war was winding down, Favo began to hold back intelligence from his boss. "I believed he was liable to say anything to Greg Scarpa," he later testified. On August 31, Favo waited until the last minute to tell DeVecchio that Scarpa Sr. was about to surrender to the NYPD on a gun charge—at which point he'd be arrested by FBI agents from DeVecchio's own squad on murder-conspiracy charges.

Favo told Dannen that DeVecchio became "visibly upset" by the news, and tried to alert Scarpa Sr. But it was too late to stop the arrest.[26]

DeVecchio's loyalty to the mad-dog killer apparently never wavered. He went so far as to try and intercede in Scarpa Sr.'s bail hearing, in an effort to keep him on the street. ASUA Andrew Weissman, who wasn't aware at the time of Scarpa Sr.'s informant status, was reportedly "incredulous" at the agent's support for the mobster. Even Scarpa Sr.'s own lawyer at the time, Joseph Benfante, was in the dark about his client's twisted relationship with Lin DeVecchio.

"That would be tantamount to me thinking that Mother Teresa [the Bureau] is assisting Saddam Hussein [Scarpa]. Because no FBI agent goes out and engages in a Columbo war. It's insanity," Benfante said.[27]

In fact, Scarpa Sr.'s status as a Top Echelon informant was concealed from three separate federal judges during the course of five separate trials.[28] Only years later, after dozens of Columbo soldiers on both sides were jailed in the war, did word of this "insane" relationship between the Supervisory Special Agent and the hitman began to leak out.

In May 1994, Anthony "Chuckie" Russo, a capo in the Colombo family, was facing a life sentence for murder and racketeering. On appeal he hired attorneys Futerfas and Resnick.

"We first discovered that Scarpa was working for the government in his sentencing transcript," says Futerfas. "He's before Judge [Jack B.] Weinstein and he says 'Judge, I helped the best I could.' That was our first hint. Then Ellen and I started reading—every piece of paper we could find; years and years of transcripts and pleadings and we started to see it. It was clear that repeatedly Scarpa Sr. had received a 'get out of jail free card' from the Feds."

Soon the two lawyers formulated a new strategy, which became known as the "comrade in arms" defense. "Much of the testimony against Russo was admitted," says Resnick, under Rule 801D2(e), the exception to the Rule against Hearsay for co-conspirator testimony.

"There were four major witnesses in Chucky Russo's trial that testified to Scarpa senior's hearsay," says Futerfas. "They said, 'Scarpa senior told me that Russo told us to go out and kill so and so. . . .' But if Scarpa wasn't a *co-conspirator* and, in fact, a government *informant,* then his testimony at Russo's trial should have been inadmissible."[29]

Two federal judges agreed, and fourteen Colombo defendants, including underboss William "Wild Bill" Cutolo, were acquitted.[30] "The floodgates were about to open," says Resnick. "We were just beginning to understand the depth of this alleged DeVecchio-Scarpa corruption, and if we had been permitted to present this theory in upcoming cases with other defense attorneys, a number of major Eastern District prosecutions could have fallen."

After one of the acquittals, a juror told the New York *Daily News:* "If the FBI's like this, society's really in trouble."[31]

But a number of Colombo bosses were jailed prior to Futerfas's discovery of the DeVecchio/Scarpa "marriage." Vic "Little Vic" Orena Jr. and Pasquale Amato both got life in prison largely on the strength of two factors: a bag of guns tied to a Colombo War homicide that was found under the deck of Vic's girlfriend's house on Long Island; and the testimony of a key witness against the two capos—Supervisory Special Agent R. Lindley DeVecchio.

Flora Edwards is a veteran defense attorney who represented Orena and Amato in 2004. "DeVecchio testified at the trial and explained the Columbo War and the hierarchy of the family and the factions," says Edwards. "But he left out a few salient little details. The first one was that Gregory Scarpa Sr. was his rat. The second thing he left out was that Gregory Scarpa Sr., was prosecuting this war almost single-handedly and killing people with DeVecchio's knowledge and assistance."[32]

In challenging their convictions, Edwards was shocked to discover from Greg Scarpa Jr. that his younger brother Joey had planted the bag of guns under Orena's girlfriend's deck while his father and DeVecchio were present in a nearby car.[33] "Here's an FBI supervisory special agent and he's an accomplice to Orena being framed," says Edwards. "Amazing."

Ultimately, she says, "the truth started to come out. There were a host of cases that followed. And every time the jury heard this evidence, that

Scarpa senior was working for the government, guess what happened? They acquitted."

Gregory Scarpa Sr. had lived a violent life, and in the end he suffered mightily. During a 1992 hospital stay to have his stomach removed, he contracted the HIV virus via a blood transfusion from one of his crew members, who later died of AIDS. By the spring of 1994, at the age of 65, as he lay dying in the prison hospital in Rochester, Minnesota, Scarpa decided to try to atone for his three-decade murder spree.

In a deathbed confession, he admitted that he had been responsible for the so-called Colombo war. He even took credit for a number of murders for which other family capos and soldiers were doing life. "But the record was incomplete," says defense attorney Edwards. "His youngest son, Joey, was murdered in 1995. So the best living witness to corroborate Scarpa senior's corrupt relationship with DeVecchio became his son. Greg Jr."

Flora Edwards arranged to have Greg Jr. testify about his father and DeVecchio via a video hookup from prison in a hearing for Orena and Amato on January 7, 2004.[34] Once again, Judge Jack B. Weinstein was presiding.

"At some point," said Edwards, "personnel in the Eastern District and the Justice Department came to understand the significance of Scarpa junior's testimony. He had to be discredited. Because once Greg Jr. became credible about Ramzi Yousef, then he'd be credible about what he was going to say in my hearing too."

Edwards pointed out that the success of the Colombo War cases in the Eastern District had helped make the reputations of more than a few prosecutors, in the same way Rudolph Giuliani's prosecution of the Mafia "Commission" cases in the Southern District of New York had made his. "These cases were career builders," says Edwards. "And if it turned out that Scarpa junior was correct and DeVecchio was not an overzealous agent but a murderous accomplice, then they would have had a real problem on their hands. Because if all of those related cases started to unravel, the careers of a lot of Justice Department officials would be impacted."

In the course of his video testimony before Judge Weinstein on January 7, the younger Scarpa confirmed what he had already sworn to in an affidavit: that his father had ordered him to make payments to DeVecchio out of their "numbers racket;" that "over the years," one hundred thousand dollars went to DeVecchio; that DeVecchio, whom Greg Jr., knew as "D," served as a lookout on a Queens bank robbery in the 1980s; that he arranged for his father to get probation after Secret Service agents sold him between fifteen and twenty thousand dollars' worth of counterfeit blank credit cards in a sting at the Wimpy Boys Social Club; and that Greg Jr., his brother Joey, and Greg Sr.'s common-law wife, Linda, even watched FBI surveillance videos of the social club furnished by DeVecchio.[35]

In dramatic live video testimony in which he appeared handcuffed with the cuffs chained to a belly belt, the younger Scarpa swore under oath that his father had killed at least two Colombo associates, Nicholas "Nicky Black" Grancio and Vincent Fusaro, after DeVecchio furnished Scarpa Sr. with their locations. He even told how his murderous father ordered him to reward DeVecchio by renting a room for him in a Staten Island hotel and supplying "a couple of call girls."

In the course of his testimony Greg Jr. also touched on his "attempt . . . to assist the government," by furnishing the FBI with information on "terrorists." But he gave no details. At the time of the testimony, defense attorney, Flora Edwards, was not aware of the depth or quality of Greg's intelligence from Yousef. Until now, that material has remained sealed in FBI and Justice Department files.

In the end, Judge Weinstein rejected Greg Jr.'s testimony. "The court finds this witness to be not credible," Weinstein wrote in a four-page decision, denying Edwards's motion for a new trial.[36]

Fearing the Case Might Unravel

"They did an ends-justify-the means decision," says Edwards. "They made a choice. Scarpa Jr. had to be made not credible."

A few weeks after 9/11, in a hearing before the same judge, attorney David Schoen, representing Colombo member Michael Sessa, was petitioning for FBI files he believed would prove that Sessa was "100% innocent of the murder for which he was convicted." Among the witnesses whom Schoen was seeking to call before Judge Weinstein was retired Supervisory Special Agent R. Lindley DeVecchio.

While granting part of Schoen's motion to obtain certain discovery material, the judge stated that he was "disinclined to go through Agent Delvecchio [sic] again." In denying Schoen's request to call the former agent, Weinstein said, "There's a time to end all of these cases. I will not permit all of these related cases to unravel."[37]

But in shielding DeVecchio, there was a problem. How could the Feds explain the extraordinary leaks of information to Scarpa Sr. that had been confirmed in multiple interviews with Colombo family members who had turned? And if Lin DeVecchio had been sanctioned in his dealings with Scarpa at the highest levels of the FBI, as he testified, then who was to blame for his malfeasance?

By early 1994, Christopher Favo and two other agents in the C-10 squad brought what amounted to charges against DeVecchio. The Bureau opened an OPR internal affairs investigation. The timing couldn't have been worse for Valerie Caproni, chief of the Eastern District's Criminal Division. Her office was then preparing to go to trial against Anthony "Chuckie" Russo and a series of co-defendants in a major Colombo war prosecution before Judge Charles P. Sifton.

"So what Valerie did," says Futerfas "was to push the Justice Department to delay the OPR. But worse than that, when Judge Sifton later demanded that the prosecutors turn over all documents relating to the DeVecchio-Scarpa investigation, she failed to turn over a key 302 involving a confession by Scarpa Sr.'s underling Larry Mazza."

Known as the "girlfriend 302," the memo memorialized a confession in which Larry Mazza admitted that Greg Sr. had a law enforcement source that "supplied information on a regular basis."* The memo went on: "The information that SCARPA SR. received through his source(s) included, but was not limited to, the address of VICTOR ORENA'S girlfriend's home."

"Understand the significance of this," says defense attorney Ellen Resnick. "It's evidence that Scarpa senior, while acting with the advise and consent of a senior FBI agent, is getting the addresses of his rivals so that

*See full 302 memo on p. 314.

he can go off and kill them. The production of that 302 to defense attorneys at a critical time could have been devastating,"

So where was that crucial document?* It remained in the drawer of Valerie Caproni's desk.[38] And Caproni went even further. As the FBI geared up to conduct its OPR on DeVecchio, she called the investigating agent and asked him to hold off.

"Caproni advised," wrote OPR Special Agent Thomas Fuentes, "that no FBI interviews of La Cosa Nostra (LCN) members expected to testify in an upcoming trial should be conducted because of the potential adverse impact on the prosecution."[39]

"In other words," says defense attorney Edwards, "further corroboration by the Colombo turncoats of the Scarpa-DeVecchio 'marriage' could result in an acquittal."

It is unlikely that Caproni's request for a slowdown in the OPR on DeVecchio was motivated by any suspicion that "Del" was innocent. The very next day after that phone call to Fuentes, the OPR special agent interviewed her again. In a five-page 302, Caproni admitted that as far back as 1987 she had known that Scarpa Sr. was an FBI informant. She admitted that information she got from DeVecchio on Scarpa's crew was of "of little prosecutive value," and that DeVecchio had threatened "to get" Agent Chris Favo if an OPR was opened on him.[40]

Futerfas and Resnick later filed a twenty-page memorandum effectively accusing the U.S. Attorney's office of obstructing justice.

Finally, after years of litigation, Judge Sifton issued a sweeping decision reversing the conviction of Russo and other Colombo defendants. Futerfas and Resnik's client would get a new trial.

But soon the tide began to turn in DeVecchio's favor. Chris Favo, who had first accused his boss, began to regret coming forward. He concluded that informing on DeVecchio with the other two agents "was a mistake that would follow us for our careers," and eventually the leadership in the FBI's New York office folded as well.[41]

In April 1996, James Kallstrom sent a memo to FBI Director Louis J. Freeh marked "Precedence: IMMEDIATE." Drafted by Jim Roth, the

*See diagram, p. 314.

principal legal officer (PLO) in the NYO, the memo was entitled "Supervisory Special Agent R. Lindley Del Vecchio [sic] OPR Matter."*

"NY requests that whatever investigation is to be conducted as a result of this letter be conducted expeditiously," Roth wrote on Kallstrom's behalf. "The failure of the DOJ . . . to administratively resolve this matter continues to have a serious negative impact on the government's prosecutions of various LCN figures in the EDNY and casts a cloud over the NYO."

That was it: the smoking gun memo that connected the DeVecchio-Scarpa scandal to the multitude of Colombo "war" prosecutions.

When it came to the DeVecchio OPR, James Kallstrom, the assistant FBI director in charge of the New York office, was effectively admonishing Washington to "shut it down."

Despite the sworn testimony of four cooperating Colombo family witnesses, including Larry Mazza, Scarpa's personal protégée; despite the charges brought by Agents Favo, Leadbetter, and Tomlinson, that DeVecchio had crossed the line and repeatedly supplied intelligence to a "Mad Dog" killer—despite all of this, the FBI closed the OPR. DeVecchio, who refused to undergo a polygraph, took the Fifth Amendment and answered forty-four questions with the words "I don't recall," was allowed to retire with a full pension.

The Cover Up

Three months later, by July 18, the DeVecchio scandal was eclipsed by the TWA 800 crash. Valerie Caproni, whose Eastern District office included the Long Island crash site, got as caught up in the incipient criminal probe as did James Kallstrom. But the linchpin between the two investigations remained Gregory Scarpa Jr.

On one hand, Scarpa was sitting in jail, awaiting trial in Caproni's office on RICO charges—waiting for a chance to become one of DeVecchio's chief accusers as he prepared to tell the details of "Del's" fourteen-year career as his father's own "informant" inside the FBI.

*See diagram, p. 315.

On the other hand, the younger Scarpa was supplying the FBI with extraordinary intelligence straight from Ramzi Yousef, the World Trade Center bomber who had threatened to cause a mistrial in the Bojinka case by getting a bomb planted aboard a U.S. airliner.

The government faced a critical choice: to embrace Greg Scarpa Jr., or to cut him loose; to reveal the extraordinary leads he had received in his contact with Yousef, or to bury them in order to keep a string of mob cases from falling apart. The choice would be made at the highest levels of the Justice Department.

And in the end—as happened at so many other points along the road to 9/11—America's national security would be sacrificed in favor of political convenience.

Faced with choosing the war on terror over organized crime, the FBI and the Justice Department decided to go with the mob.

8

THE FORTY-YEAR REWARD

As they sought to plug the holes in the sinking ship that the Colombo War cases had become, prosecutors in the Eastern District saw Gregory Scarpa Jr. as a problem. The younger Scarpa had kept the books on his father's numbers racket, and was prepared to say that Greg Sr. had used the bookie slush fund to pay off DeVecchio. Further, the younger Scarpa was willing to testify that he had *personally* benefited from DeVecchio's leaks to his father. In fact, he'd gone on the lam in 1987, after his name appeared on prosecutor Valerie Caproni's arrest list.

By 1995, Greg was preparing for his own RICO trial, and it was clear what his defense would be—that his father had been responsible for the crimes that he was being accused of, and that the FBI, via DeVecchio, had sanctioned his behavior.

DeVecchio, meanwhile, denied the charges vociferously. His lawyer, Douglas Grover, called them "scandalous and ridiculous." In May Grover appealed directly to FBI director Louis Freeh, complaining that "the investigation of Special Agent DeVecchio has 'kicked around' on 'the slow track' for over one year. Each delay in its resolution has provided fuel for defense attorneys in significant organized crime prosecutions in the Eastern

District of New York."[1] Grover demanded a "quick and decisive resolution of the investigation."

In truth, though, his client had the Bureau over a barrel. If the younger Scarpa was believed and the "comrades in arms" defense continued to flourish, prison cells would open on a dangerous crew of mobsters who were clearly guilty of other offenses.

One of the Justice Department officials copied on that letter from Grover was Jamie Gorelick,* the deputy attorney general. A few months later, on August 22, Gorelick would join Louis Freeh among the DOJ officials who huddled with James Kallstrom in that fateful Washington, D.C., meeting that produced the FBI's about-face in the TWA 800 investigation.

In the spring 1996, Justice Department officials like Gorelick had faced a difficult choice: either admit that the Colombo War cases were hopelessly tainted and cut their losses—with any number of convictions getting tossed—or ride it out and clear DeVecchio by seeking to diminish the credibility of witnesses like Greg Scarpa Jr.

"So they rolled over," says Flora Edwards, who sought unsuccessfully to use Greg Jr.'s testimony years later. "Some senior people in Justice just *acquiesced.* The evidence that DeVecchio was providing illegal information to Scarpa was absolutely overwhelming," she says. "But so was the perception in the New York judiciary that you do not go easy on the mob."

Defense attorney Alan Futerfas agrees. "There was a belief in this city for years," he says, "that the Mafia controlled construction, the fish industry and the waste industry. People like Rudy Giuliani made their careers by breaking that perceived control. Now, the Colombos constituted a large series of prosecutions. And no matter what we presented in the way of government misconduct—obstruction of justice and withholding of evidence—the appellate court was not going to let these men out of jail. When the Court of Appeals reversed Judge Sifton, who had granted the Russos new trials, the implicit message was that the ends justified the means—the conduct may be reprehensible, but the convictions stand."

*See picture, p. 268.

482 Months in Solitary

In the fall of 1998, cut loose by the Feds, his remarkable intelligence on Yousef disparaged as "a hoax," Greg Scarpa Jr. went on trial. There was a hint of how he would fare in a September 28 *Daily News* piece by Jerry Capeci, a reporter whose career had been made with a weekly column called "Gang Land," which appeared every Tuesday. Under the headline "My Father Did It," Capeci wrote, "by blaming his father, Scarpa seems to be copying tactics that 16 other Colombos used to win acquittals a few years ago when they went to trial."[2] Capeci was critical; his story had a pro-Bureau slant. But what Capeci didn't tell his readers was that one of his top sources over the years had been Lin DeVecchio.

In an FBI 302 memo dated June 1–2, 1994, "Del's" number two, Special Agent Christopher Favo, confessed that "on several occasions, SSA DeVecchio approached me or another agent with specific questions about the investigation of the war or the Colombo family and our answers to the questions appeared in Capece's [sic] column the following day with an attribution to a law enforcement source.[3] On several occasions I mentioned this to SSA DeVecchio and he blamed the leaks on the prosecutors in the EDNY. I began withholding information from SSA DeVecchio on Monday mornings to prevent its appearance in Tuesday's column."

In another column, at the start of Greg Jr's. defense at trial,[4] Capeci mocked him again: "Colombo mobster Gregory Scarpa Jr. has probably been reading a lot of cloak and dagger thrillers," he wrote. "He opened his defense to racketeering and murder charges with a heavy dose of international intrigue, espionage and double cross."

After describing Greg's testimony about the spy camera he used to photograph Yousef's kites, Capeci cracked that "Scarpa said in many ways he was like his father . . . a top echelon FBI informant for 30 years who regarded himself as a James Bond figure. *Mission Impossible* was a favorite television show."

True to his source, Capeci dutifully noted that "DeVecchio, testifying under a grant of immunity from prosecution, denied any wrongdoing from the witness stand." Capeci went on to describe how Linda Diana Schiro, Scarpa Sr.'s girlfriend, who was supposed to testify for his son, "took the Fifth" after prosecutors told her "she could be grilled about crimes she was allegedly involved in."

Capeci also reported that "Faced with the same problem, Scarpa's fall-back witness, his mother Connie, also refused to testify."

In the story, headlined "I Spy," Capeci disparaged the quality of the intelligence Greg Jr., had gathered from Yousef, and noted that Judge Reena Raggi had rejected Junior's motion to preclude rebuttal testimony. What the columnist *didn't* mention was Judge Raggi's history with Yousef—specifically, her ruling in a crucial case related to Yousef and his WTC bombing co-conspirator Mohammed Ajaj.

Sending Bomb Manuals Back to a Terrorist

Ajaj was the figure who had flown into JFK with Yousef on September 1, 1992, and served as his cover as he sought political asylum. Ajaj, a Palestinian, presented an INS agent with a Swedish passport that was an obvious fake. After agents found a half dozen fake passports and bomb manuals in his suitcase, Ajaj got the last bed in an INS detention facility. Amid the distraction, the wily Yousef was allowed to slip into the country unchallenged.

In December 1992, as Yousef was building the 1,500-pound device he would soon plant beneath World Trade Center's north tower, he spoke regularly to Ajaj, then in the U.S. penitentiary at Otisville, New York. The three-way calls were patched through the burger restaurant of Ajaj's uncle Abu Omar in Texas. In one of their coded conversations—which prison officials apparently failed to note at the time—Yousef requested that Ajaj contact a judge to get the bomb manuals back so that he could refer to one of them as he built the complicated urea-nitrate device.[5]

Naturally, federal prosecutors objected to Ajaj retrieving such material, especially since they had checked out his background with Israeli intelligence and learned that Ajaj was a member of the notorious al-Fatah terrorist organization.[6] The Feds wanted Ajaj to get an *upward* departure date—a longer prison stay.

And yet Judge Raggi had other ideas. On December 22 she told an AUSA: "This defendant pleaded guilty in October and these documents have been in the government's possession since September. I see no reason why this has been delayed . . . terrorism constitutes . . . a very serious allegation, and, in this case, I would have to be persuaded by a preponderance of the evidence . . . that the defendant, when he entered the United States, entered using the false passport in an attempt to further terrorist action."[7]

Astonishingly, Raggi then proceeded to grant Ajaj's motion to get back the contents of his suitcase—which included "books and manuals on how to use hand grenades, how to commit sabotage, how to make poisons and Molotov cocktails, how to place land mines, instructional materials on how to kill with a knife, diagrams depicting how to make silencers, videotapes about suicide car bombings and how to make TNT."[8]

On December 29, 1992, barely two months before Yousef bombed the Trade Center, Ajaj called him to report gleefully on his legal victory. Speaking on a pay phone outside of his Jersey City bomb factory, Yousef asked to see one of the bomb manuals. But Ajaj, speaking in code, noted that it wouldn't be a good idea for Yousef himself to retrieve the volatile material, because it might jeopardize his "business," which would be "a pity."[9]

Now, four years later, on October 23, 1998,[10] Greg Scarpa Jr.—the man who had been the FBI's secret weapon against Ramzi Yousef—was convicted of various RICO violations . . . and the judge who would pronounce sentence, Reena Raggi, was the same woman who had seen fit to reunite Yousef's accomplice with his manuals before the WTC bombing. Surely she had learned a lesson about the dangers of dismissing a terrorist threat lightly—or had she?

Given her personal experience with the Ajaj case, Judge Raggi was in a better position than most Eastern District federal judges to appreciate the growing al Qaeda threat. And this was no abstract fear: by the fall of 1998, Federal officials had linked Osama bin Laden and al Qaeda to the bombings of the U.S. embassies in Kenya and Tanzania that past August. In September FBI agents from the SDNY had arrested Ali Mohammed, the ex-Egyptian army officer-turned Green Beret who had run the Calverton shooting sessions where members of Ramzi Yousef's bombing cell were trained back in 1989. Mohammed was accused of personally surveilling the African embassies for bin Laden.

In that same month—October 1998—the Justice Department also indicted Wadih El-Hage, the Arizona-based al Qaeda operative who had been bin Laden's personal secretary in the Sudan. The indictment laid bare bin Laden's ties to the murderous Islamic Group (IG), which had been run by Sheikh Rahman—another link to the Yousef cell. And within a month of the start of Greg Scarpa Jr.'s trial, the Feds unsealed its indictment of bin Laden himself—a kind of road map to the al Qaeda network based on the testimony of Jamal al-Fadl, who had become a secret FBI witness in 1996. It was Al-Fadl, the jihadi, who, once and for all, provided the link between bin Laden, al Qaeda and the Yousef-Rahman cell.[11]

If any judge in the EDNY had the wherewithal to appreciate the quality of Greg Jr.'s intelligence on Ramzi Yousef, then, it was Reena Raggi. If she believed in the quality of Greg Jr.'s intelligence, Judge Raggi might have granted his 5K1.1 motion and rewarded him for risking his life by shaving some time off his sentence. Larry Silverman, the mobster's lawyer, was hoping that the "substantial assistance" he'd given the Feds would garner Greg a significant downward departure date.

Instead, Judge Raggi sentenced him to forty years.

As a point of comparison, Sammy "the Bull" Gravano, the Gotti underboss who admitted to nineteen murders, received a prison sentence of five years.[12] Greg Scarpa Jr., who was *acquitted* of five murder charges, got 482 months in solitary. Worse, he was ordered to do his time in the Administrative Maximum Security (ADMAX) Penitentiary in Florence, Colorado, a subterranean penitentiary that is considered the toughest prison in the U.S. federal system. Also known as Supermax, it's home to some of the most lethal criminals ever convicted—including Terry Nichols, twice convicted in the Oklahoma City bombing, and Unabomber Ted Kaczynski. Today, the prison also houses Ramzi Yousef himself—after drawing a 240-year sentence from Judge Duffy in 1998 following his conviction in both the Bojinka and World Trade Center bombing trials, the master bomber was relocated to the same facility as the mobster who once conspired with the FBI to bring him down.

"It's an absolute outrage that Greg Scarpa Jr. should be doing four decades in the same jail as the man behind the Trade Center bombing and Bojinka," says Silverman. "Especially when we only *now* have a full understanding of the quality of intelligence he was furnishing to the Feds."

The Quality of the Yousef-Scarpa Intelligence

An examination of Scarpa Jr.'s handwritten kites, and the FBI 302s that authenticate them, leaves little doubt that the intelligence Scarpa procured from al Qaeda's chief bomb maker could have thwarted the TWA 800 crash and allowed the FBI to capture Khalid Shaikh Mohammed as he mounted the 9/11 plot.

There's no question that the intel from Yousef was genuine. Not only did Ramzi's May 19, 1996, bomb formula match precisely with the details of his Casio-nitro bomb trigger—a design unknown to *anyone* outside of U.S. and Philippines intelligence—but Scarpa got details from Yousef about Ramzi's capture in Islamabad on February 7, 1995, that could only have come from the master bomber himself.

In an FBI 302 on September 19, 1996, twelve days after his conviction in the Bojinka case, Yousef told Scarpa about Istaique Parker, the young South African Muslim he had recruited in 1995 to help him plant bombs aboard a series of United and Delta flights leaving Asia. In the months since, Parker's name had turned up in only a few select U.S. publications as the informant who ultimately betrayed Yousef in return for a $2 million reward the State Department was offering for his capture.

Now, according to a September 19, 1996, FBI 302 quoting Scarpa, Yousef revealed that he had trusted Parker to plant the bombs aboard flights leaving Thailand, but after the young South African saw a "Time or Newsweek" with Yousef's picture describing the reward, Parker decided to betray him.[13]

No one, certainly not a member of the Colombo crime family, could have known such details without getting them *directly* from Ramzi Yousef himself.

That report was one of thirty-seven separate intelligence briefings that Greg Scarpa Jr. gave the Feds over the course of eleven months between March 5, 1996, and February 7, 1997. One of the FBI 302s, from December 30, 1996, revealed a threat by the bomb maker that if he wasn't freed, Osama bin Laden, aka "Bojinga," would order the hijacking of airliners to free Yousef's fellow al Qaeda cell members, along with blind Sheikh Omar Abdel Rahman and the other nine defendants convicted with the Sheikh in 1995 in the "Day of Terror" plot to blow up the bridges and tunnels around Manhattan.

That very prediction—an airplane hijack plot to free the Sheikh—was contained in a 1998 Presidential Daily Briefing (PDB) sent to President Clinton that was one of the news-making revelations after publication of the 9/11 Commission's final report. And the identical threat was included four years and eight months later in the controversial August 6, 2001 PDB* received by George W. Bush at his ranch in Crawford, Texas.

"We have not been able to corroborate some of the more sensational threat reporting," said the controversial memo, "such as that from a [redacted] service in 1998 saying that Bin Laden wanted to hijack a US aircraft to gain the release of 'Blind Shaykh' Umar Abd al-Rahman and other US-held extremists."

The threat warnings leading up to the 9/11 attacks contained in that single PDB were considered so important by the 9/11 Commissioners that the document almost provoked a constitutional crisis when the Bush administration moved unsuccessfully to keep it classified.

Now, for the first time, we have proof that the FBI was in possession of that same threat information, directly from Ramzi Yousef, in 1996—two years earlier than the earliest such intelligence cited by the 9/11 Commission.

In an appendix to this book are reproduced selected examples of the dozens of kites Scarpa provided to the FBI, along with the 302 memos that confirm their authenticity. All of the documents are available for analysis on our web site at www.peterlance.com. After examining them, it seems clear that no one with a detailed understanding of Ramzi Yousef's

* See PDB on p. 308.

bomb-making methods and post-WTC bombing terror plots could conclude that the material from Scarpa was anything but genuine.

Indeed, *Daily News* reporter Greg Smith, who first broke the story of the Scarpa-Yousef relationship, reported in a September 24, 2000, story that Patrick Fitzgerald, the head of the Organized Crime and Terrorism Unit of the SDNY, submitted a sealed affidavit on June 25, 1998, in which he said that a follow up investigation "appeared to corroborate Scarpa's information."[14]

But it didn't seem to matter. Lin DeVecchio was cleared in the FBI's internal OPR and retired on a full pension. Greg Jr.'s Yousef information was called "a hoax," and he got forty years.

The other appeals based on the "comrades in arms" defense failed. The FBI sided with the NTSB in the TWA 800 investigation, and the apparent cover-up held—that is, until a courageous young forensic intelligence analyst decided to help a cop she believed had been set up by the Feds to take the fall for DeVecchio.

"Nothing Equaled the FBI Misconduct in This Case"

Angela Clemente* is a 39-year-old single mother of three. With one adult daughter, another teenage daughter still at home, and a young son born with a life-threatening allergic disability, she works as a freelance researcher for a series of defense attorneys and constitutional lawyers.

"It's mostly pro bono," says Clemente, whose main motivation as a righter of wrongs is government reform. Clemente, who had always been interested in forensics, began her career as an assistant in a pathology lab. In 1991, after learning of a particularly egregious rape case involving a repeat sex offender, she began volunteering to get the laws changed in Florida, where she lived. "This was a precursor to Megan's Law," she says. "We were shocked that repeat sex offenders could travel freely and there was no way to keep track of them."

* See picture, p. 266.

Then in 1997 she became a crime victim herself. On the night of July 19, she was raped by a man in Seattle. "I just crawled into a hole after that," she said. "I could barely function for a couple of years."

It was another desire to change the law that sent her back to work.

"My brother is a prison guard," says Clemente. "A few years ago, while working in the Midwest, he observed some other guards abusing an inmate with a diminished mental capacity. He reported them to his superiors and found himself shunned. He was shocked. I was shocked. So I started making some calls. We ended up getting some real reforms in that state prison hospital."

After that victory, Clemente began receiving letters from other inmates, asking for help with their appeals. Soon, she found herself doing research to overturn some convictions she believed were unjust. That led to a professional association with Dr. Stephen Dresch, a Yale Ph.D. and political economist who had served a term in the Michigan House of Representatives.

Before long, the two of them became aware of the defense work being done by Alan Futerfas and Ellen Resnick to expose the Scarpa-DeVecchio scandal. "Nothing we had ever encountered equaled the FBI and Justice Department misconduct in this case," said Clemente. "The most shocking part of it, when we began lining up the dots, was the connection to TWA Flight 800. There was a point in the summer of '96—after the explosives they found on the plane confirmed what Greg Jr., was saying—that Justice officials had to decide, 'Do we use this information—which would mean crowning Greg Jr. as our star witness in any upcoming trial against Yousef for TWA 800—or do we cut him loose, discredit him, call all that amazing material he got from Yousef worthless, a hoax?'" Angela answers the question this way: "All you have to do is look at how the FBI did a complete turnaround and closed the TWA case to see which way they went."

"The problem is," says Dresch, "they didn't just engage in a cover-up. To make it work they had to blame somebody for the Columbo leaks—so they literally set out to target a veteran New York cop. The FBI didn't just take a pass on Ramzi Yousef's New York terror cell—they also tried to ruin an honest man."[15]

9

AN NYPD COP TAKES THE FALL

The Scarpa-DeVecchio investigation led Angela Clemente and Dr. Dresch to the case of Detective Joseph Simone,* a decorated New York City police detective who had worked on the Columbo Task Force as part of a joint NYPD-FBI organized crime unit called the OCID (Organized Crime Investigation Division).

"Of all the victims stemming from this unholy alliance between an FBI agent and a mob killer, one of the saddest living victims is Joe Simone," says Clemente. "When the FBI needed an explanation for how so much intelligence had leaked to Scarpa Sr. during the Columbo wars and concluded that cases would fall if they couldn't move against the real source—DeVecchio—they had to find a fall guy, and that was Joe."[1]

Detective Simone was born in the Grave's End section of Brooklyn, an Italian neighborhood in the St. Simone Jude parish around West 6 Street and Stillwell Avenue.

"I was brought up in Flatbush," he says. "We were a close-knit Italian family—meals together on Sunday and everything. When I was six we

* See picture, p. 266.

moved to Flatbush. We lived on a block where you were either a cop or a fireman or you got made."[2]

Simone's father died when he was still a teenager, leading him to his first brush with the mob. "There was a guy name Louie who lived across the street," he recalls. "He approaches me and says, 'All you have to do for me is drive and you can make seventeen thousand bucks a year.' That was real money back then. It was at a point when we were gonna lose the house. But when I told my mother this guy had made me this offer she gave me a crack across the face and said, 'Don't get involved with those people.' So I didn't."

Eventually Joe's uncle Ernie, a first grade detective with the NYPD, convinced him to take the exam for the police department, and pretty soon he was walking a beat. In his early career in the 70th precinct, Simone was decorated multiple times.

In January 1980, as part of the 70th Precinct's Anti Crime Unit, Joe was credited with saving the life of a 74-year-old woman he rescued from a burning building in Flatbush.[3] The following fall Simone, his partner, and two other uniforms took down a pair of sexual predators wanted in thirty-six rape cases.[4]

"The whole city was after these guys," says Joe. "They had been doing break-ins, rapes, and robberies across four precincts. My partner Pat Pesce and I were on a routine patrol in Flatbush-Midwood when we spotted this silver-and-black '79 Caprice that one of the victims remembered. We staked out the car, and when two people came back they led us to the perps."

Simone and Officer Pat Maggiore* made another spectacular arrest involving a serial rapist, burglar, and arsonist who would break into homes, sexually assault the female occupants, and then set fire to their houses.[5]

"This guy was really sick," says Joe. "If the women had kids, he would burn their baby clothes. If they had fish in a tank, he would cut off the fish heads."

Realizing that the home invader was gaining access to the buildings using the Brighton elevated subway that ran behind them, Simone and his

* With officers Brian Waterboury and Sgt. Dennis Haug.

partner staked it out and, after a backyard chase, collared the suspect, who had burn blisters on the palms and fingers of each hand.

Simone's success won him a transfer to Manhattan South Narcotics; he later worked Brooklyn until 1986. There he was able to use his Grave's End street sense as a successful undercover narcotics investigator, making hundreds of dangerous buy-busts and finally earning his gold detective's shield.[6]

While working in narcotics Joe had always worn his hair long, with a full beard. Years later, after he was asked to joint the elite OCID unit and began working in the FBI's buttoned-down New York office at 26 Federal Plaza, he trimmed it back into a more conservative goatee. But the Columbo wiseguys he mixed with always called him "The Beard."

"I knew these guys and they knew me," Simone recalls. "They knew what side of the street I walked on. There was an association, but I never crossed the line. Once when my mother was in the hospital, Wild Bill Cutolo, a Colombo boss, got word she was sick. I got a call from one of his gophers who asked me if there was anything they could do. I said, 'Thanks. No. Just prayers.' The next day this kid from a florist came with a basket maybe four feet wide and three feet high with a card [saying] 'Best wishes to your mom,' signed 'Billy Cutulo and friends.' And my mother, who was as honest as the Virgin Mary said, kidding, "I hope I don't find a horse's head under my bed."

All told, Simone and his partner were involved in 33 of the almost 100 arrests during the two-year Colombo war. He was considered one of the most aggressive officers in the unit.[7]

Simone's immediate boss in the joint NYPD-FBI Division was Christopher Favo, DeVecchio's number two. An Italian American Notre Dame graduate, Favo came to New York, according to Simone, with little or no sense of the street.

"I helped to educate him," he says. "Chris knew that I knew a lot of people in the street—a lot of wiseguys. One time he came up to me and he had a picture. He says, 'Do you know this guy?' And I say, 'Yeah that's Bobby Attanasio.' He asks me, "How do you know him?' And I say, 'His kid and mine played touch football together. At the time he was a soldier in the Bonnano family. His brother was a capo. He's doin' time right now. He just got locked up in that big ring where they pulled in all the Bonnano guys.'

"Another time Favo shows me a picture—'Who's this guy?' I say, 'Joey Ida. He lives about eight blocks away from me. He's Jimmy's brother.' Then Favo asks me, 'Is he a made guy?' I say, 'Yeah. His brother's a capo and he ended up being a street boss. Now he's doing three life sentences.'"

The whole purpose of joint task forces like OCID and the NYPD-FBI Joint Terrorist Task Force was to pair educated but relatively inexperienced agents like Favo with street-smart cops like Simone. It was Joe's job to "work the wiseguys," to mix with them and gather intelligence. "I wasn't 'under' like Joe Pistone,"* says Simone.

"The Colombo people knew I was a cop, but they just felt more comfortable telling me things because I was a neighborhood guy and knew how they operated. There's no way they would have trusted some agent from the Midwest just 'cause he was Italian."

Above Favo, Simone's ultimate boss in the unit was Lin DeVecchio, another Italian-American who was a stranger to New York. The son of an Army colonel, DeVecchio was born in Fresno, California.

"Lin was a nice guy," recalls Simone. "He talked to me quite often. I had my own opinions that he was involved—he was told to let go of Scarpa several times and he never did. I knew that he was seeing Scarpa. [Chris] Favo was mentioning it. But I never expected him to be Scarpa senior's 'girlfriend,' even though there was a law enforcement leak."

Unfortunately for Simone, it was a leak the Feds had decided to plug. It is unclear who made the call, but at some point Simone was targeted by the FBI to take the hit. The Feds waited until the day he was due to retire from the NYPD after nineteen years.

The Morning the Roof Caved In

Years earlier, Detective Simone had hurt his back in the line of duty, and he was due to retire with a three-quarter tax-free pension—the holy grail for cops and firefighters in the City of New York. But in the early morning hours of December 8, 1993, a series of police and FBI vehicles pulled up

* The FBI agent who infiltrated the Bonnano crime family as "Donnie Brasco."

outside his modest house in Staten Island where he'd lived for thirteen years with his wife, Eileen, a nurse, and their five children. Within hours, Simone was under arrest for bribery.

Agents of his own OCID unit had accused him of being the source of leaks to the Columbo wiseguys. A two-decade career was flushed in an instant. The next day, rather than celebrating at a retirement party, Simone's name was smeared across the tabloids.

"Detective Stung by Feds" was the banner headline in the New York *Daily News*. "Detective Joseph Simone, who worked on the NYPD-FBI Organized Crime Task Force for seven years, was charged with selling information to the Orena faction of the warring Colombo family for two years, earning at least $2,000," the article said.[8]

The author of the piece was Jerry Capeci, the *Daily News* columnist later cited in FBI 302s as Lin DeVecchio's source for leaked stories. But the public didn't know that.

Overnight, Joseph Simone went from a hero in his Staten Island community to a tabloid disgrace. "The roof caved in on my life," said Simone. "I never saw it coming."

For years Joe had been active with his kids in the Little League and Tottenville High School football on Staten Island. Occasionally he visited the home of the school's coach, Phil Ciadella, whose uncle Alfonso "Chips" DeCostanza was a capo in the Columbo crime family.

"We had locked him up for guns during the 'war,'" remembers Joe. "He used to inform for us. I'd go over there once in a while 'cause Phil's mother cooked old-style Italian.[9]

"So one day I go to Phil's mother's house and who's there but Big Sal Micciotta, and Bo Bo Malpeso, two Columbo capos. I was shocked. . . . I didn't expect them to be there. I just went there to pick up plays. The kids were going to away to summer camp for football. So these guys started talking to me like, 'if you could help us out. . . .'"

Simone says the mob capos were feeling him out, to see if he would be willing to sell them information. "Bo Bo tried to pass me a piece of paper, but I didn't touch it," he recalls. "I couldn't tell you to this day if it was a shopping list or an envelope." Simone had lived by street rules since his days as a kid in Grave's End, and he knew a set up when he saw one.

"I told 'em, 'You guys are probably wired. I don't want any part of you.'" At that point Malpeso stripped down to his pants to show he wasn't wearing a recording device. But as Joe left, he went eyeball to eyeball with Phil Ciadella.

"I told him, 'you got some fuckin' pair of balls puttin' me in a situation like this,'" Simone remembers. "Not long after that, I informed Favo and DeVecchio of what had happened."

It was after a meeting with Favo that Simone believes the Feds decided to set him up formally. In July 1993 he was called back to the Ciadella house by the coach.

Court records show that in the interim Micciotta had agreed to inform for the Bureau. Two agents later claimed that he approached them. But suspiciously, there was no paperwork on file indicating that he'd made the offer.[10]

Moreover, Micciotta alleged that in the "paper" Bo Bo Malpeso had proffered to Simone, there was $1500 in bribe money. Big Sal claimed that Simone took it, but the Feds needed corroboration. So a second meeting was arranged.

This time Micciotta was carrying a tape recorder. A transcript of a conversation he had before Simone arrived suggests that Micciotta wasn't very optimistic about being able to corrupt the veteran cop.[11] The tape begins with Big Sal wondering to DeCostanza whether he can get "help" from Simone after getting "pinched again."

> MICCIOTTA: I need a little information. I wanna find out what the fuck they're gonna do with this case here. . . . You know, I'd do the right thing. I'll take care of him.
>
> DECOSTANZA: He don't want nothin'.
>
> MICCIOTTA: I'm glad to give him a coupla dollars.
>
> DECOSTANZA: You don't have to give him nothi' . . .

Moments later, in a reference to the flowers that Wild Bill Cutolo had send to Simone's mother, Big Sal says:

MICCIOTTA: Billy's the guy who burned him out, he sent him flowers. . . . In other words, he reached out for the guy; the guy didn't respond.

"It's clear from the transcript of this tape in which Micciotta is wired, that he believes that Detective Simone would not take a bribe," says Angela Clemente, the forensic investigator who worked Joe's case.

Then, for unexplained reasons, just prior to Simone arriving, Micciotta shut off the tape recorder.

During the meeting, Big Sal confessed to Simone that his son was "on the lam down in Florida." The younger Micciotta was wanted by the NYPD's 6th Precinct for attacking a young seminarian, but it was his father who had broken the priest-in-training's arm.

"As soon as I saw Favo again," remembers Simone, "I told him about the son. 'I got it from the horse's mouth,' I said, that he was down in Florida. I told him that we ought to notify the Precinct Detectives Squad where he is."

After briefing Favo, Simone thought nothing else of the meeting. It was his job to connect with mob guys and get information. He'd told Favo, his immediate FBI supervisor, about both encounters. And then he went about his job, looking forward to retirement as the summer and fall of 1993 passed.

Short FBI Tape May Aid "Rogue Cop" Defense

Simone's arrest came just as his family and friends were preparing to throw him a retirement party. By the end of the day he was handcuffed and charged with taking a bribe from Micciotta. With the tape off, it came down to the word of a decorated veteran cop against that of a mobster with an interest in cooperating with the Feds. But Simone lost.

Despite the fact that he had agreed to talk to cops and FBI agents for three hours after his arrest without a lawyer present, and despite the fact that he offered to take a polygraph—which the Feds declined—Detective Joe Simone was indicted.

"Initially they had sixty counts against me," says Simone. "Everybody was trying to dump what went bad in the Colombo wars on me—like wires . . . giving up CI's. . . . Everything. They ended up coming down to four counts: two attempted alleged bribery and two attempted alleged conspiracies."

In the *Daily News,* though, Jerry Capeci and reporter Tom Robbins virtually had him convicted: "the FBI learned of Simone's turncoat role last May from a Colombo soldier who agreed to wear a wire and later taped Simone and DeCostanza in several incriminating conversations."[12]

The story never mentioned that the tape had been shut off and there was no record of the alleged bribe encounter. The piece contained no qualifiers. The word "alleged" was never used; the story was printed as fact. The article noted that "DeCostanza (whose cooperation the Feds sought) won a dismissal of weapons charges arising from an arrest during the height of the Columbo war."

There wasn't a word of speculation as to whether that might have been the payoff to "Chips" from the FBI for setting up the veteran detective. The story noted that Simone had now been suspended from the NYPD and released on a $50,000 personal recognizance bond along with DeConstanza.

Two days later, Capeci and Robbins weighed in with a story about "Big Sal" Micciotta. In 60 point type, the headline read "Mob biggie aids FBI sting."[13] The piece, which repeated the story that Micciotta had been wired, noted that he had been "hustled into federal protection" for his role in the arrest of "the gang's top-secret mole in FBI headquarters—a New York City Police Detective."

By April, new evidence emerged that should have produced a retraction from Capeci. The FBI admitted that Micciotta had switched off his tape recorder during his meeting with Simone, so there was no record of the alleged bribe that the *Daily News* had reported. Undeterred and unwilling to admit a mistake, Capeci filed a story under the headline, "Short FBI Tape May Aid 'Rogue Cop' Defense,"[14]

"Ever since he was charged with selling his badge to the mob," wrote Capeci, "Detective Joseph Simone has waged a fierce and desperate battle for his reputation, his job, his pension and his freedom. . . . And thanks to a screw-up in the FBI plan to trap the suspected rogue cop, the double agent defense may work for him."

There was no mention in the piece about Simone's heretofore unblemished nineteen-year career, or the fact it was his word versus that of Big Sal, a violent Columbo killer who had mercilessly brutalized a young priest.[15]

Micciotta later lied on the witness stand in the murder-racketeering trial of six other Columbo members, and was dropped by the Feds from witness protection. But Capeci never backed down in his reference to Joe Simone as a "suspected rogue cop."

Meanwhile, as hamhanded as the Feds had been in their attempt to set up Simone on the bribery charge, they were even more inept in trying to frame him.

At trial in the same Brooklyn courthouse where Greg Scarpa Jr., had been indicted, they presented evidence alleging that Simone had used his desk phone at 26 Federal Plaza to signal Columbo members (Scarpa Sr., style), punching a series of 6's into a beeper. They also alleged that a female FBI agent on a nearby payphone outside a Staten Island deli had overheard him attempting to contact a known mob associate.

On the day of his arrest the FBI had cleaned out Simone's desk. As far as federal prosecutors knew, the detective had no records to support his defense.

"But what they didn't count on," says Joe, "was that I had my DARs."

DARs were Daily Activity Reports, which chronicled Simone's service to the OCID in quarter-hour increments going back for years. Simone had saved them all in an inch-thick file.

"Every time they had me at work tipping off one of these made guys, the DARs proved that either my tour of duty was over or I was on vacation," said Simone, brandishing the pink DARs that convinced the federal jury of the FBI frame-up.

"One time they actually used my office phone to beep a wiseguy, claiming I was trying to tip him off with the 666—and it turned out that I was in Wildwood, New Jersey, at the time with my family. No way was I gonna drive all the way back to Federal Plaza, three and a half hours, dial some 6's, and then go back down to Wildwood."

"Later, they said I was using a pay phone outside a deli to tip off the wiseguys, and they claimed to have a female FBI agent on an adjacent phone listening in." At trial, FBI Special Agent Lynn Smith testified that on September 23, 1993, she was part of an eight-person FBI surveillance team following Simone. Under oath she alleged that they had tracked him to a deli-superette on the corner of Arthur Kill Road and Elverton Avenue, and that she had overheard him talking on a phone "located outside [the] door of the superette." Asked by Joe's defense lawyer John Patton if there had been one phone or two at the location, she replied, "One on the left and one on the right, on each side of the door."[17]

"But as it turned out there was only one phone there," Simone says. "We had the deli owner who was prepared to testify that there was only one phone on the side of the building that I used and there never had been two phones by the front door where this agent said she'd been."

"Understand the significance of this," says Angela Clemente. "The FBI is so intent on nailing this poor cop that they put eight agents on him. A female agent testifies under oath to a phone that doesn't exist so they can make it look like Joe is calling to tip off wiseguys. The truth is, he used to stop at that deli before coming home to call his wife and ask if she wanted him to bring home cold cuts for dinner. But the Bureau goes to all this trouble to make it look like he's caught up in some major leak to the Colombos."

In another instance, defense attorney John Patton showed the jury a series of documents supplied by the FBI, which purported to prove he'd taken $1,500. But the lawyer, a veteran of defending cops, lined up the staple marks on the documents to prove that one of the incriminating pages had been replaced by the feds *after* Joe's arrest.[18]

One of Simone's chief accusers at trial was his FBI case agent, Christopher Favo. What the jury didn't realize at the time, though, was that for almost two years leading up to Simone's arrest in December 1993, Favo had suspected *his* immediate supervisor, Lin DeVecchio, as the source of the leaks. But he'd kept quiet about his suspicions until January 1994, a month after Simone's arrest, when Favo and agents Leadbetter and Tomlinson came forward. In the months that followed, Favo poured out his concerns about Scarpa Sr.'s allegedly tainted relationship with DeVecchio in a series of FBI 302s.[19] But none of that kept him from taking the stand against Joe Simone.

"At the start of trial, I was worried," recalls Simone. "Not because I was guilty, but because of the way the jury was made up. There were these twelve Nordic, blue eyed, blond-haired people," he said. "There wasn't an Italian-American among them except for an alternate. No blacks. No Hispanics. None of the people you think of as New Yorkers. And I'm sittin' there scared shitless. I see these people starin' at me—especially the foreman. He just stared me down during the whole trial."

A Break from "Mr. Organized Crime"

Ironically, the real turn in the proceedings came one day during a lunch break when Simone's brother John was reading the *Daily News.* "There had been a huge DEA bust," remembers Joe. "And my brother calls me over with the paper and says, 'Do you know any of these guys?' And I looked at it and it said that Lindley DeVecchio was in charge of this *DEA* Task Force. I'm thinking, that can't be right. He's Mr. Organized Crime. So I go up to John Patton and say, 'this doesn't make sense—DeVecchio's supposed to be the top guy in the country on wiseguys. Why would he be working drugs?'"

Neither Patton nor his client had any idea of the reason for the switch: That by then, under FBI suspicion, DeVecchio had been moved over to a DEA position. Though an OPR had been opened on the accused agent, aside from a few rumors Simone had heard, nobody outside of the Bureau or the Justice Department knew that DeVecchio was a target of the FBI's internal affairs unit.

"So right after lunch, when Favo gets on the stand," Simone recalls, "John Patton decides to play a hunch and he takes a shot. 'Is it true,' he asks, 'that your supervisor DeVecchio is under investigation for corruption?' Suddenly, you can hear a pin drop in the courtroom and Favo says, 'I'm not allowed to disclose that at this time.'

"That was it," says Simone. "Confirmation that something was wrong. John turned around and gave me the Groucho eyebrows. That's how we discovered that Lin DeVecchio was the bad guy and I had been set up to take the heat off him."

But the Feds—who by then were well aware of DeVecchio's leaks to Scarpa—didn't make it any easier on Simone and his lawyer. "We tried several times to get the phone records," Patton recalled. "What are called LUDS—records from Joe's phone—that we could compare against his DARs. And each time, mysteriously, for those particular weeks, the phone people told us they were missing."

"You have to understand how outrageous this was," says Patton. "Sal Micciotta was the personification of evil. The guy had no moral center. And it was simply *his* word against Joe's without a shred of probative corroborating evidence. The fact that they even got an indictment was outrageous. These prosecutors ruined the life of a great cop just to cover for a dirty agent."

Ultimately, though, after hearing all the evidence this jury that had worried Simone came back before the end of their lunch break. "John looked at me and said, 'this is either very, very good or very, very bad,' " he remembers. "But they acquitted me on all counts. The foreman I was scared of stood up and yelled out, *'Not guilty!'* I cried like a baby."

10

THE DEATH OF NICKY BLACK

The *New York Times* ran a prominent two-column story: "Detective Is Found Not Guilty of Selling Secrets to the Mafia."[1] *New York Newsday* did the same: "Cop Not Guilty of Fed Rap."[2] Joe's hometown paper, the *Staten Island Advance,* gave him a banner headline: "Detective Acquitted of Mob Charges."[3] Later, they followed up with a piece headlined "Cop puts his life back together. Simone: 'It feels good to be free.'"[4]

But Jerry Capeci's byline didn't appear in the *Daily News* story on the acquittal. That job was left to Greg B. Smith, the respected crime reporter who'd been the first journalist to publish word of the Yousef-Scarpa intelligence six years later. In a short single-column piece headlined "Juries show little respect for 'Big Sal,'" Smith reported that "A Brooklyn federal jury took just two hours to declare Detective Joseph Simone innocent." Even that vindication was buried in the third paragraph of a story that focused on the extraordinary efforts by the Feds to protect the murderer and priest-beater Salvatore Micciotta. Still, Smith quoted a juror who told him that "no one believed the Feds' witnesses, including Micciotta, who claimed Simone had fed him secrets."

Unrepentant, two weeks later Jerry Capeci published a Gang Land column headlined "Cop still treading hot water."[5]

"He was acquitted of federal charges that he sold his badge to mobsters during the bloody Columbo war," wrote Capeci, "but it may not yet be all over for Detective Joe Simone. . . . The Feds can't simply call a 'do over' in Simone's case. But they can—and are—pushing city cops to pick up the ball they dropped and prosecute Simone on departmental charges that would cost the 19-year veteran his pension."

"Despite the acquittal charges, Gang Land sources say FBI agent Chris Favo and assistant U.S. Attorneys Stanley Okula and Karen Popp firmly believe Simone was guilty as charged." Quoting an unnamed cop, Capeci blamed the acquittal on "just a bad jury."

Once again, the *Daily News*'s readership was left in the dark. Because of the secretive nature of the OPR process, no one outside of the FBI and Justice Department knew that one of the charges DeVecchio would face in his OPR was Christopher Favo's accusation that the agent had regularly leaked confidential FBI information to Jerry Capeci.

The only hint of Simone's point of view in the Gang Land piece was a quote from his lawyer John Patton: " 'We're not afraid of a G.O. 15,' Patton said, in a reference to the regulations that force cops to answer all questions or be fired."

A Second Prosecution for Detective Simone

Sure enough, the NYPD's Internal Affairs Division went ahead and filed charges. "They essentially mirrored the federal indictment," said Patton, who was confident of an acquittal, especially after several FBI agents who were former colleagues of Simone's agreed to testify in his behalf.

But the NYPD "indictment" included a charge that Detective Simone had "wrongfully failed and neglected to report to this Department a bribe offer." That reference was to the piece of paper that Joe had refused to touch, and which the lying Micciotta had described as an envelope containing $1,500 in cash.

Again, the proceedings would hinge on the word of a cop, whom the hearing examiner admitted had been "hand-picked" for the elite Columbo Task Force, against the allegations of one of the very suspects he had investigated: Big Sal Micciotta.

This time, in the NYPD trial, Chris Favo was determined not to get caught off guard. When asked if his immediate supervisor DeVecchio was under investigation for leaking information, Favo said that he had been "directed not to answer."[6]

In the end, despite the evidence that Joe's work rating was consistently designated "exceeds standards," the hearing officer came down with what Patton described as "a horrendous decision." Even though it was part of Simone's job description to meet with mobsters and elicit information for the OCID task force, Deputy NYPD Commissioner Rae Downes Koshetz (who tried the case without a jury) decided that the paper allegedly offered to Joe *had* been a bribe and that he should have reported it to his police department superiors. It didn't seem to matter that he had told both Chris Favo and Lin DeVecchio about the encounter with Big Sal.

Further, Commissioner Koshetz seemed to conclude that Simone's "guilt" was also directly related to his association with Phil Ciadella, the Pee Wee football coach, whose uncle, "Chips" DeCostanza, was in the mob.

Joe had testified fondly that he enjoyed going over to Phil's house, because Ciadella's elderly mother and father "would treat me like their own son." He described them as "seventy-, eighty-year-old people that I had respect for." Despite the fact that neither of those pensioners had a criminal record, Commissioner Koshetz righteously declared: "Members of this Department are forbidden to associate with known criminals. Even if Ciadella was not on a database, his connection to the mob was too close for comfort. Moreover, the Respondent's claim that associating with such people is part and parcel of living and raising a family in certain parts of Staten Island, is unavailing; if a New York City police officer can not conduct his personal life without associating with mobsters or their close relatives, he is expected to move elsewhere."

"This astonishing decision in which Det. Simone was found guilty and stripped of his pension, suggested one of two things," said Patton. "Either the hearing officer was incredibly naïve with respect to the fact that Joe was working in an O.C. squad and expected to mix with wiseguys as part of his job, or she was doing the bidding of the Feds."

One hint of an answer to that question came when William J. Bratton, who was police commissioner at the time of Simone's guilty verdict, put off signing the final papers that would have stripped Joe of his pension.

Bratton, now the chief of the LAPD, eventually retired and was replaced by then-FDNY Commissioner Howard Safir.

Safir had spent thirty-five years as a Fed, serving as assistant DEA director before becoming chief of the U.S. Marshal's Witness Protection Program—a unit with intimate ties to the Southern and Eastern Districts, which monitored many of the turncoat mobsters that the FBI's New York office sought to protect.

Safir was sworn in as NYPD Commissioner on April 15, 1996, three days after Koshetz's verdict against Simone. Eleven days later attorney Patton sent him a six-page heartfelt appeal letter[7] requesting a reevaluation of the verdict. In it Patton cited the testimony of the FBI agents and NYPD superior officers who had testified on Joe's behalf. He summarized the flimsy evidence against Simone and underscored what was by then twenty-two years of loyal service to the Department.

But Safir was unmoved. On May 5, Patton was notified that Joe had been terminated. He would leave the NYPD without a pension.

"In any case," says Patton, "Joe was set up. They should have given him a chance to get out and take care of his five kids."

The blow to his family and his reputation sent Simone into a spiral. "First there was severe depression, then I wound up becoming an alcoholic," he says.

Simone was bitter. "At one point during the trial, Favo had this smirk on his face," he remembers. "And I went up to him and I said, 'You little shit. You didn't cheat *me* out of my pension. You cheated my wife and my kids, and I will never forgive you for that.'"

Now Simone, the veteran cop who put a third of the wiseguys away during the Colombo war, resorts to working odd jobs. His wife Eileen works full-time as a nurse to make ends meet. Having gone through rehab, he now attends seven twelve-step meetings a week as he tries to piece his life together, enjoy his grandchildren, and forget.

R. Lindley DeVecchio, on the other hand, was cleared of any charges of wrongdoing. The OPR on him was closed, and he retired with a full FBI pension.

And the story might have ended there, if Angela Clemente and Dr. Stephen Dresch hadn't kept digging.

The Violent Death of "Nicky Black"

Looking back over the evidence that documented Lin "Del" DeVecchio's years as the "Grim Reaper's" handler, the two legal investigators found new evidence to explain why the Feds seemed to go out of their way to target Detective Joe Simone.

"Of course, they needed somebody to sacrifice in order to explain the leaks," said Clemente. "But that wasn't enough to explain why they actually framed Joe. Then, as we began peeling back the layers, we realized that Detective Simone was in a position to become one of the lead witnesses against DeVecchio if the Feds had ever brought him up on charges. And they had reason to. It all stemmed from the murder of Nicholas Grancio."

The murder Clemente referred to had occurred on the afternoon of Monday, January 7, 1992. The Colombo "war" that Scarpa Sr. had instigated was at its height. Wiseguys of both the Persico and Orena factions drove around Brooklyn and Staten Island "strapped"—armed to the teeth. "There was a sense that a gun battle could break out at any moment," recalls Simone.

At the time, Simone was assigned by the FBI to follow Nicholas "Nicky Black" Grancio, a Colombo capo loyal to the Orenas whose nickname derived from the dark circles under his eyes.

"He had a new white Toyota Land Cruiser," recalls Simone, "and my partner Pat Maggiore and I followed him to outside a social club he had on McDonald Avenue in Brooklyn."

The OCID Colombo Task Force, which had a "perch," or surveillance post, designated as "Plant 26" on the second story of a building overlooking the site, was keeping almost round the clock surveillance on Grancio.

On the day before, a Sunday, Simone and his partner followed Grancio and one Ralphie Piccirillo to the plant, trailing the Land Cruiser in a white FBI Nissan Maxima. At 11:20 A.M., when they pulled up to the social club, the two cops hung back, then went upstairs to the perch, where they could get out of the January cold and monitor Grancio with surveillance equipment.[8]

"Nicky sat right in front of the door to this plant," remembers Simone. "The entrance was on Grave's Neck, Avenue U. There was a store on the bottom. Upstairs we had the camera and recording equipment." The two cops stayed there that day until 4:00 P.M., when their tour ended. "It was uneventful," says Simone.

What he and his partner didn't know then was that Greg Scarpa Sr. was out gunning for Grancio, checking the mob social clubs all over Brooklyn for him. The next day, Monday, when Simone and Maggiore were back in a car outside Plant 26 watching Grancio's Toyota, Scarpa drove past the site in a dark van.

Greg Sr.'s number two, Larry Mazza, later testified that the driver was another member of Scarpa's crew named Jimmy Del Masto. Scarpa was in the back with a shotgun.

Defense attorney Flora Edwards gained some insight into the Scarpa-Mazza relationship while interviewing Mazza in 2002 when he was on supervised probation after pleading guilty to a series of murders he committed with the elder Scarpa. "Larry was like a second son to Greg senior," says Edwards. Mazza "opened up" to her in the interview, she says, which took place at the Hilton Hotel in the Orlando Airport.

When the dark van approached the white Toyota, Mazza told Edwards, he and Scarpa realized the Feds were watching. "They wanted to kill Grancio," she says, "but they couldn't get a clear shot at him because the cops were all over him."[9]

"So what happened then," says Edwards, "is that Scarpa senior borrowed Mazza's cell phone and called somebody named 'Del.' He said, 'What the fuck is going on here? The whole world's here. Do something."

Moments later, at 2:00 P.M. (14:00, as it would read on police records), Detectives Simone and Maggiore and the other Feds watching Grancio got a strange call. "Out of the clear blue," says Joe, "Favo calls and tells us all to come back to Federal Plaza for a Team Meeting. Now this is unusual, since it's two in the afternoon and we normally meet at the end of the day."[10]

As outlined in Simone's DAR, reproduced on page 315, the FBI surveillance team then withdrew and headed back to 26 Federal Plaza for the meeting with Favo at 4:00 P.M. (16:00).

"About ten minutes after they left for 26 Federal, there wasn't a cop in sight," says Edwards. "All the surveillance was gone. So Larry told me that he and Scarpa senior pulled up next to the SUV where Grancio was sitting and blew his head off."

Minutes later, the meeting at the FBI's New York office was interrupted with word that Grancio has been murdered.

"So we take off," says Joe, "and rush back there. At the crime scene we find that the back of Nicky's head's been blown away.

"Joey Tolino, a Columbo soldier who was leaning against the Toyota talking to him on the passenger side, later tells me he saw this van pull up, the door roll open, and [the] barrel of this rifle he thought was a shotgun. Then *blam*. . . . He ended up with brain matter all over him. We even recovered one of Nicky's teeth from the wall of a house fifteen feet away. It was gruesome."

Later, when Simone interviewed Tolino at a nearby precinct, the wiseguy wanted to know what had happened to the Feds who had been watching over his faction in the war. "He said to me, 'Man, you guys were following us all this time, and all of a sudden you disappear and they whack Nicky,' " says Joe. "I didn't have the heart to tell him at the time that we got pulled back for a team meeting in the Ivory Tower at 26 Federal."

If the call from Scarpa did go to DeVecchio, and he was responsible for Favo's "unusual" request to pull back the team, that could make DeVecchio—a senior FBI agent—an accessory to homicide.

On May 6, 1993, at a sentencing hearing before Judge B. Weinstein in the Eastern District, Greg Scarpa Sr. pleaded guilty to the Grancio murder.[11] Later, Larry Mazza pleaded guilty as an accomplice.

Mazza not only told attorney Edwards about the Grancio hit, he also told government reform investigator Dr. Stephen Dresch, who maintains a web site, www.forensic-intelligence.org. In 2003, when Dresch himself interviewed Mazza—this time in Plantation, Florida—he came away with the identical story that Mazza told Flora Edwards. Only this time Mazza was even more specific.

"He told me that he was with Scarpa senior when the crew went to whack Nicky Grancio," says Dr. Dresch. "Finding Grancio under surveillance by the organized crime task force, including Joe Simone, Scarpa called DeVecchio, who had the task force team pulled off the surveillance."[12]

At a January 7, 2004, hearing on the twelfth anniversary of the Grancio murder, Dresch testified under oath about his conversation the year before with Mazza, declaring that after "Scarpa called . . . his law enforcement source," Scarpa and Mazza "returned to the scene, discovered that the surveillance team had been terminated and then . . . proceeded to terminate Mr. Grancio."

The hearing was in front of Judge Jack Weinstein, the same judge who in 1993 had accepted Scarpa Sr.'s guilty plea for the Grancio hit.

But when it came time for Flora Edwards to ask Larry Mazza to swear under oath to what he'd told her and Dr. Dresch, the confessed killer got cold feet. In an interview for this book, Edwards recounted her conversation with Mazza this way:

> EDWARDS: Gee, Larry, will you give me an affidavit?
>
> MAZZA: No.
>
> EDWARDS: Will you testify to this in court?
>
> MAZZA: No.
>
> EDWARDS: Why, Larry? 'Cause you already pled to the Grancio murder. You're not hurting yourself.
>
> MAZZA: 'Cause I'm afraid of the FBI . . . the government. They're gonna trump something up. I'm gonna end up back in jail if I'm lucky, and if I'm not lucky, they'll kill me.

"When you add up all of these pieces, you begin to understand what the stakes were here for the government," says Angela Clemente, who has spent the last two years, working largely pro bono, building a file on this case. "You can see why they had to set up Detective Simone. His DARs proved that he and the surveillance team got pulled away by Favo, who's working directly for DeVecchio. This was confirmed by Flora Edwards, a prominent defense attorney, and Dr. Stephen Dresch, a Yale Ph.D."

Clemente says now that she fully understands the Feds' motive in targeting Simone. "This scandal involving DeVecchio was monumental for the FBI's New York office. If the full depth of his corruption with Scarpa senior had come out, not only would these Colombo convictions have unraveled, but a top Supervisory Special Agent might have stood accused as a murder conspirator. Joe Simone had to be disgraced to diminish his role as a witness in all of this, and Greg Scarpa junior had to be taken out of the picture as well."

Clemente, who personally unearthed the Yousef-Scarpa kites and confirming FBI 302s, doesn't mince words when it comes to Greg Scarpa Jr. "He may have done other bad things in his life," she says. "He was a member of the Mafia. He sold drugs and engaged in other kinds of racketeering. But on the issue of his credibility with respect to Ramzi Yousef, after months spent examining that material, I believe he was one hundred per cent truthful."

The 9/11 Commission Ignores the Evidence

In April 2004, hoping that someone in official Washington would pay attention to their findings, Clemente and Dresch brought their evidence to the House Government Reform Committee.* They also sent a time line detailing their findings to the 9/11 Commission.[13] While they had several tentative Capitol Hill meetings with the Reform Committee staff, they have been ignored by the Commission.

"One would think that evidence linking Ramzi Yousef to the crash of TWA Flight 800 might be relevant to the Commission," she says. "It's clear now that Yousef and his uncle [Khalid Shaikh Mohammed] were involved in both attacks on the World Trade Center, but since we submitted our material to the Commission [on] April 5, we've heard nothing."

And therein lies the key: By ignoring the research of independent investigators like Angela Clemente and Dr. Stephen Dresch, the federal government is continuing its pattern of covering up the truth about the long road of negligence that led inexorably to 9/11.

*See letter, pp. 303–4.

At first blush, it may seem difficult to accept that so random an event as an obscure Brooklyn mob hit could have been one of the signposts along that road. Yet consider: If the FBI hadn't been afraid to expose the involvement of one of its own agents, R. Lindley DeVecchio, in that murder, Gregory Scarpa Jr., might never have been unfairly discredited as a witness. And if Scarpa's evidence of Ramzi Yousef's involvement in the TWA 800 crash had been allowed in court—and pursued aggressively by the FBI—Yousef's hijacked-airliners suicide plot might have been identified years before its culmination in 9/11, with its mastermind Khalid Shaikh Mohammed brought to ground.

From DeVecchio to Scarpa to Yousef to KSM: Only four degrees of separation stood between the FBI and the nascent 9/11 plot—five years before it was set in motion.

Yet none of this has ever been addressed, at least in public, by the 9/11 Commission. That body's disregard for Clemente's and Dresch's findings only corroborates what we have learned in this investigation: that evidence linking Ramzi Yousef to 9/11 has been systematically excluded from the Commission's carefully censored account of the attacks. In the second half of this book, we'll examine how the Commission—charged with getting to the truth about the greatest act of terror on American soil—has failed in its mission.

PART II

My kids went to school. They were nine, seven, and three. I walked them to the bus stop. My son, who was three, was skipping home, and I'm on the phone making a play date for him, and the girl I was talking to said, "I can't believe it. I'm watching TV and a plane just hit the Trade Center." And within two minutes of that call my mother-in-law, who had lost a brother at nineteen, came in screaming "it's happening again . . ."

—Mindy Kleinberg, whose husband Alan, thirty-nine, was a trader with Cantor Fitzgerald on the 100th floor of the North Tower

I was home exercising and the phone rang. But I didn't get to it in time, and that will always haunt me. When I played the message, it was my husband, Kenneth. He said, "I'm in the World Trade Center. The building was hit by something. I hope I'm going to get out. But I love you very much and I hope I'll see you later." The next thing I know I'm in the living room in a fetal position, hysterical, screaming.

—Lorie van Auken, whose husband Kenneth, forty-seven, worked for Cantor Fitzgerald on the 105th floor of the North Tower

I was rushing to take our daughter to speech therapy when my husband called. He said, "Sweets, I'm fine," and I'm like, "Okay . . ." not realizing, and I said, "Call me later. I'm running late." And he said, "I didn't want you to worry. Put the television on." So I walked into the kitchen and put on the little TV, on and I thought Oh my God. And he said, "I was sitting at my desk and all of a sudden I looked over and there's this huge fireball." And he goes, "Sweets, you don't understand, people are jumping out of the windows . . ."

—Kristen Breitweiser, whose husband Ronald, thirty-nine, worked for Fiduciary Trust International in the South Tower

I was doing a wash when John called after the first plane had hit. He told me that the stairwells were gone and the roof doors were locked. He said the floor was filling with thick black smoke and he didn't think they were going to make it. And he broke down and sobbed. He told me over and over, "I love you. I love you. Tell John"—our son—"that I love him." He was a very devout Catholic—he had a lot of faith—and I told him that if anyone was prepared, he was. Then I told him how much I loved him. I can't remember how the call ended—if we were cut off—but from there I went into shock.

—Patty Casazza, whose husband John, thirty-eight, was a bond trader for Cantor Fitzgerald on the 104th floor of the North Tower

11

THE WHITE HOUSE STONEWALL

On that black Tuesday, four women from New Jersey who had never met heard the news that would alter their lives. Within days each of them, in her own way, began trying to find answers to the many questions they all had in common. "We simply wanted to know why our husbands were killed," says Kristen. "Why they went to work one day and didn't come back."[1]

The four women were strangers before that day. Two of them had voted for George Bush in 2000, and two had voted for Al Gore.[2] But Kristen, who, first traveled to Washington with the others, early in 2002, was clear at the time about her motivation. "It's not about politics," she said. "It's about the safety of the nation."[3]

"The Congressional Joint Inquiry was really limited in its scope," says Patty of the earlier governmental investigation that examined the attacks. "Whole areas of investigation were off limits, like the day of 9/11 and defense. It was beyond their scope to investigate how four airliners could be hijacked and used as missiles, and how NORAD and the FAA just seemed to let it happen. We needed to know why."

"The Joint Inquiry only had six months to cover 30 years of intelligence failures," says Mindy. "And even then, when their report came out in July [2003], it was heavily redacted."

Of the four and one half pages devoted to Khalid Shaikh Mohammed in the Joint Inquiry's report, more than half were blocked out for security reasons. This underscores how little information was given to the general public about the man the FBI has pegged as the driving force behind 9/11.

"We wanted an investigation, based on a legal model," says Lorie. "In a negligence case you have a finding of fault and you have a damage phase. The Victims Compensation fund, for all of its limitations, was dealing with the damages. We wanted an investigation with teeth—subpoena power, the ability of the investigators to grill witnesses under oath—a probe that wouldn't be subject to the same political considerations you'd expect from an inquiry up on Capitol Hill."

Breitweiser, an attorney who had practiced law for only three days before becoming a full-time wife and mother, began using her skills as an investigator. Meanwhile, across the Hudson River in New York, Monica Gabrielle was forming the Skyscraper Safety Campaign with Sally Regenhard, whose son Christian, a probationary firefighter, was one of the 343 FDNY victims of the attacks. Monica's husband, Rich, worked for Aon Corporation on the 103d floor of the South Tower. She later learned from survivors that he lay injured on the 78th floor lobby, unable to move or get word out to her, before the building collapsed. So Monica and Sally began demanding answers.

"We wanted to know why two one hundred and ten-story skyscrapers could fall in little over one hundred minutes from the moment of the first crash," said Regenhard, "and why occupants of the South Tower had been told that it was safe for them to remain in the building."

"We found ourselves asking a lot of the same questions as Kristen and Lorie," says Monica. "So eventually we joined forces." Together with the Jersey Girls and other survivors like Carol Ashley of Rockville Center, Long Island, whose 25-year-old daughter, Janice, died on the 93d floor of the North Tower, they began traveling to Washington to push for an independent investigation.*

* See list of the Family Steering Committee membership, p. 316.

By late January 2002, Sen. Joseph Lieberman (D-CT) was co-sponsoring a bill to set up what eventually became the 9/11 Commission. But both President Bush and Vice President Cheney personally lobbied Senate majority leader Tom Daschle to back off.[4]

"The vice president expressed the concern that a review of what happened on September 11 would take resources and personnel away from the . . . war on terrorism," Daschle told reporters. "But clearly I think the American people are entitled to know what happened and why."

It took another nine months for the widows to lobby the Commission into being. On November 27, 2002, President Bush signed HR 4628, the Intelligence Authorization Act for Fiscal Year 2003, which established the body formally known as the National Commission on Terrorist Attacks Upon the United States.*

"The Commission will build upon the work of the congressional joint inquiries," said the president,[5] "to carefully examine the circumstances surrounding the attacks and the lessons to be learned from them. I expect that the Commission's final report will contain important recommendations for steps that can be taken to improve our preparedness for and response to terrorist attacks in the future."

"But almost from the beginning, it was clear," says Sally Regenhard, "that this administration was going to fight this panel tooth and nail."[6]

The first clash between the White House and the families came shortly after the bill's signing, when President Bush announced his appointments for Commission chairman and co-chairman. Former Nixon Administration secretary of state Henry Kissinger was chosen to lead the investigation, along with former Maine Democratic Senator George Mitchell.

"The White House invited us to the signing and they brought Kissinger out with Mitchell," says Lorie. "We watched it on TV and were shocked. Kissinger had huge conflicts of interest—major dealings with the Saudis."

Days later, Kissinger resigned after refusing to disclose his client list.[7] The Jersey Girls could take some of the credit for his hasty departure.

"The day before he resigned," says Lorie, "we had a meeting with him in his office in Manhattan.[8] Kristen had done impeccable research. She'd looked up all of his companies. So I asked him, 'Mr. Kissinger, do you

* See pictures of Commission members, p. 268.

have any Saudi clients?' He mumbled something. And then he asked if someone would pour him some coffee. So then I said, 'Do you happen to have any clients by the name of *bin Laden?*' He almost fell off the couch."

Kissinger and Mitchell were soon replaced by Tom Kean, a moderate Republican who'd served two terms as New Jersey governor, and former Democratic Congressman Lee Hamilton, who had chaired the House Intelligence Committee. The remaining eight Commissioners, four from each side of the aisle, were virtually all Beltway insiders, with the exception of former Illinois Governor James Thompson.[9]

"Once the Commission was set up formally, we started asking questions," says Breitweiser. "Like what about the FBI? How did they get the pictures of the nineteen hijackers so soon after the event? Had they been tracking them? What about the allegations of insider trading—that there had been unusually high put options taken in the stock of American and United airlines in the days prior to the attacks, betting that the stock would go down? Who had prior knowledge of the attacks? And what about the relationship of the Saudis to the Bush family?

"These were issues we felt needed to be examined."

Philip Zelikow and Dr. Rice

But the next major hurdle for the Commission was funding. By the spring of 2003, after an initial startup appropriation of only $3 million, the White House was stalling on a crucial $11 million funding request that would have kept the Commission going through its original target date for completion of its report: May 27, 2004. As a point of comparison, $50 million was earmarked for the investigation into the crash of the space shuttle Columbia, in which seven people died.[10]

"The administration was tying the Commission's hands at almost every turn," says Kyle Hence, co-chair of 9/11 Citizens Watch, an independent watchdog group. "By the summer of 2003, even the *Wall Street Journal*[11] was reporting that the Commission investigators were only getting a small portion of the documents they'd asked for from the White House. The staff had just begun conducting interviews and that was, what? Seven months after they'd been set up with an eighteen-month out date."

The White House resistance became so acute that Commission investigators were denied access to the full declassified 800-plus-page report of the Joint Inquiry until late spring, leading Sen. John McCain, one of the original sponsors of the Commission, to declare, "While I don't want to believe such a basic lack of cooperation was intentional, it nonetheless creates the appearance of bureaucratic stonewalling."[12] McCain had originally pushed for a two-year investigation. The White House demanded twelve months, and they reached a compromise at eighteen. But this didn't sit well with the families.

"When is the last time you ever heard of a homicide investigation being closed-ended?" asks Beverly Eckert,* whose husband, Sean Rooney, spoke to her on his cell phone up until minutes before he perished in the South Tower.[13] "The Commission was almost set up to fail, with all of the limitations that were placed on it in terms of budget and time."

By early July, even moderates Kean and Hamilton were expressing their exasperation. In an interim report on July 8, 2003, they complained that the "Administration underestimated the scale of the Commission's work and the full breadth of support required."[14] The report was particularly critical of the Pentagon, where document requests relating to NORAD had received no response. "The problems that have arisen so far with the DOD are becoming particularly serious," said the report. Commission staffers also complained that "agency representatives" or "minders" were present during interviews.

"How much candor can you expect from an FAA flight controller when his boss is breathing down his neck?" said a confidential source on the Commission staff. "It was a joke."[15]

"We were starting to get worried after that first report," said Mindy Kleinberg, "but by October, when we found out who *Zelikow* really was, we hit the ceiling."

*See picture, p. 263.

Zelikow was Commission staff director Phillip Zelikow,* a professor at the University of Virginia. Digging into his background, the Jersey Girls—who had now expanded into a twelve-member Family Steering Committee (FSC)—had learned that Zelikow had been appointed to President Bush's Foreign Intelligence Advisory Board in October 2001 . . . and six years earlier he'd written a book with Condoleezza Rice.[16]

"If he's looking at the NSC, that means he's investigating himself," said Lorie van Auken at the time.[17] "What an incredible conflict," said Mindy Kleinberg. "The evidence was pointing to a series of key memos in the summer of 2001 including the August 6th PDB and Condi was at the heart of that. How could we expect Zelikow to fairly investigate his own friend and colleague?"

In a letter to the Commission, the FSC demanded that Zelikow recuse himself "from any aspect of national security and executive branch negotiations and investigations." If he was unwilling to do so, they insisted, Zelikow should resign.[18]

But Kean and Hamilton rejected the FSC's demand for Zelikow's resignation, saying that Zelikow had previously agreed to recuse himself from NSC issues relating to the Clinton-Bush transition.

"That wasn't enough," said Kristen Breitweiser. "The performance of the National Security Council was a crucial issue before this Commission, and eventually, when the White House stonewalled on access to NSC documents, who ended up being one of the two commission people with full access? Philip Zelikow."

The other Commissioner who was appointed to see all documents was former Clinton Justice Department deputy attorney general Jamie Gorelick, the Commissioner who had attended the 1996 Washington meeting that led to the shutdown of the TWA 800 investigation.

But the deal on access to classified documents struck by the Commission and the White House caused an even greater firestorm in the 9/11 victims' rights community. It was struck after the White House balked on the release of "very sensitive" documents, and the Commission threatened to subpoena them.[19]

* See picture, p. 268.

"We called it the two-to-four-to-ten deal," said Patty Casazza. "Instead of allowing all ten Commissioners to go in and examine key documents like the President Daily Briefing of August 6th, they set up this system in which Zelikow and Gorelick would go in and then Kean and Hamilton would review their report, which could be edited by the White House. Then this final version would get seen by everybody."

Admitting that he went along with the compromise "because we are under tremendous time pressure," Commissioner Richard Ben-Veniste, a former Watergate prosecutor the widows had expected to be a firebrand, acquiesced. At the time he acknowledged that the deal allowed the staff "minimum acceptable access."[20]

"But that was the very problem," says Mindy Kleinberg. "The White House kept stonewalling the Commission staff. This [combined with the Committee's mandated finish date] produced the pressure that led to this compromise. I mean, we're only talking about the last official body charged with investigating almost three thousand homicides."

"At that point, in October going into November," said Monica Gabrielle, "it was clear to us that the Commission was having to drag the truth out of the White House."

The widows were further stunned when they learned that in almost a year the Commission staff had not issued a single subpoena or taken a word of testimony under oath in open session. Instead, its early public hearings had been largely "informational"; one, on Terrorism, Al Qaeda, and the Muslim World, included the testimony of a discredited scholar who was a key advocate for the false assertion leading up to the Iraqi invasion that Saddam Hussein had been tied to the 1993 WTC bombing and the 9/11 attacks.[21]

The View from inside the Commission

When I began this phase of my investigation in the fall of 2003, I developed a confidential source on the commission staff. Frustrated at the glacial pace of the Commission's probe, he began meeting with me in New York, where we would get together after work at coffee shops and restaurants in Manhattan, and he would vent his feelings.

The source, who had a heavy law enforcement background, revealed that of the eight "teams" set up to investigate various aspects of the attack by December 2003 (five months before the final report was due), only one, which he described as the "New York Team," had issued subpoenas. It was run by John Farmer, a former New Jersey attorney general who was close to chairman Tom Kean. "The other teams are completely controlled by Zelikow down in D.C.," he said.[22]

"Farmer is really butting heads with him," said the source. "Zelikow is calling the shots. He's skewing the investigation and running it his own way. What's worse, none of the other team leaders talk to the Commissioners. Farmer is the only one who deals with the Commissioners, because he has that relationship with Kean."

By the late fall of 2003, the media seemed largely uninterested in the Commission's work. "The hearing we had November 19 on Emergency Preparedness was in New Jersey," said Monica Gabrielle, who testified with Sally Regenhard. "But not one of the major New York TV stations even crossed the river to cover it. The networks were AWOL. The *New York Times* wasn't there. We were nowhere at that point."

My Commission staff source complained that time was running out. He was even considering resigning so as not to sign his name to a report that only got at part of the truth. "The Commission needs more time," he said, "and right now it looks like they're not gonna get it."

On November 27, 2003, the Family Steering Committee came out formally with a request that the Commission be granted an extension to finish its work.

"Ironically," wrote the FSC in a statement, "the production of a timely report no longer seems to be possible, in large part because of the delays caused by the Administration and the agencies who report to it. Due to the untimely issuance of subpoenas to the FAA, NORAD, and the City of New York, along with the access restrictions placed by the White House (resulting in a compromised Commission), the FSC strongly recommends that this Commission explore the possibility of an extension."[23]

The problem was that any effort to move the Commission's May 27, 2004, deadline would require the approval of both Congress and the White House. With the Bush campaign concerned about pushing back the Commission's report further into the election year, however, no such accommodation was likely.[24]

Silencing Senator Cleland

For weeks, the White House had taken a number of hits from former Georgia Senator Max Cleland, one of the ten Commissioners, who was vociferous in his criticism of the compromise over access to classified records.

"It should be a national scandal," said Cleland in an interview with Eric Boehlert for Salon.com on November 21.[25]

"It's been painfully obvious that the administration not only fought the creation of the Commission but that their objective was the war in Iraq and one of the notions it was built on was that there was a direct connection between al Qaeda and 9/11 and Saddam Hussein. There was not. I feel like I have been duped."

On the issue of full Commission access to documents like the PDB's Cleland, a triple Vietnam amputee declared that "We shouldn't be making deals. We issue subpoenas. That decision [with the White House] compromised the mission of the 9/11 Commission pure and simple." Further, he described the deal as "Nixonian" in approach. The approach should be, "Yes, the 9/11 commission gets access to the documents, all the commissioners get access. Whatever items you request we'll be forthcoming. . . . We're coming down to the final [months] of this Commission and we're still messing with access issues. The President ought to be ashamed."

Within two weeks of that interview, the *New York Times* announced that Cleland would resign from the Commission to take a position on the board of the Export-Import Bank. His name had been submitted months earlier by Senate Democrats, but it was only after his open attacks on the Bush Administration that the White House sent his nomination to the Senate.[26]

The Family Steering Committee had lost its strongest advocate for full disclosure on the Commission.

12

THE CHICKEN COOP AND THE FOX

Cleland was soon replaced by former Democratic Senator Bob Kerrey, who was then serving as president of the New School University in Manhattan. The Family Steering Committee members were cautiously hopeful. Known as a "maverick moderate-liberal," Kerrey was seen by some as the "Democrats' John McCain."[1] But Kerry's appointment did little to placate the Jersey Girls, who were now becoming openly critical of the Commission's work.

"There's a big discrepancy between what we wanted and what is happening," said Patty Casazza. "I was looking for an *investigation* into September 11, 2001," said Lorie van Auken. "That is not what we got."[2]

Mindy Kleinberg was even more bitter: "I don't know if I can stomach sitting through another hearing that doesn't get to the heart of the matter," she said. "But if we don't go, who's watching?"

At the time, in early December, there was minor support for an extension among congressman on Capitol Hill, but the White House was holding fast to the Commission's original deadline. Then moderate Republican Tom Kean, the Commission's chairman, made a surprising statement. In an interview with CBS News on December 17, he suggested that the 9/11 attacks might have been preventable.

"As you read the report," Kean said, "you're going to have a pretty clear idea what wasn't done and what should have been done. This was not something that had to happen."[3]

Even more surprising, from the mild-mannered Kean, was the suggestion that those at fault should be held accountable. "There are people that, if I was doing the job, would certainly not be in the position they were in at that time because they failed," he said. That interview provoked allusions to Watergate. The headline in one piece two days later was "What did Bush know and when did he know it?"[4]

As the new year turned, the widows began to feel that they might get support for an extension. By mid-January Michael Isikoff was reporting in *Newsweek* that "a new political battle is brewing" over whether the Commission should get an extension. "The prospect of unleashing the report in the middle of the election season is creating anxiety inside the White House."[5]

Three days later, the *New York Times* revealed that both Phillip Zelikow and Jamie Gorelick had been questioned as witnesses by their own Commission staff.[6]

"Investigators interviewed [Zelikow] to learn how much information the incoming administration had about the possibility of a major attack," said the piece, "and what steps it took to guard against that threat."

Gorelick told *Times* reporters Eric Lichtblau and James Risen that she had been interviewed about her involvement in terrorism policy while Deputy Attorney General in the Clinton administration. But the *Times* piece pointed up the conflict in which the Commission had found itself. "Mr. Zelikow's arrangement [to recuse himself] has caused concern among some commission officials because the man responsible for day to day operations of the panel will be removed from what could be an important part of its inquiry."

On one hand Zelikow was supposed to remove himself from NSC-related matters that related to his White House tenure. Yet he was one of only two Commission representatives with full access to all the files.

"By this point they should have put Zelikow's picture in the dictionary next to the word 'conflict,'" says Lorie van Auken. "I don't know how you put it any other way," says Monica Gabrielle, "Phil Zelikow was the fox guarding the chicken coop."

The issue of full disclosure became even more acute months later, when the *New York Times* broke another story revealing that the Bush White House was blocking the Commission from access to more than nine thousand pages of classified foreign policy and counterterrorism documents from the Clinton years.[7] And by mid-January 2004, when the Commission formally requested an extension from President Bush and the Congress, the White House made it clear that it opposed the idea.[8]

A few days later, the *Washington Post*'s Dan Eggen reported that the Commission would be forced to scale back hearings as it rushed to complete interviews with up to two hundred additional witnesses and examine more than two million pages of documents by late May.[9]

A Showdown with the Widows

On February 3, the widows decided to take off the gloves.

In a formal statement they asked for an extension of the Commission work for another year, and warned that "a pre-mature termination of this Commission will place this nation at risk."[10] Their press statement asserted that:

> Prior to this Commission terminating its investigation, the Commission must:
>
> • Fulfill its legislative mandate of fully investigating the attacks of 9/11 and going wherever the facts may lead;
>
> • Fulfill its verbal promises to the American public regarding public hearings;
>
> • Fulfill its promise to gain access to all relevant information regarding 9/11, including information contained in Presidential Daily Briefings and held by federal whistleblowers; and

- Fulfill its promise to conduct a transparent investigation that provides accountability and fixes responsibility to those who contributed to the failures that led to 9/11.

The next day, the media reported that the White House would agree to some kind of extension,[11] but Republican House Speaker Dennis Hastert rejected the proposal. "The worst thing that can happen to this commission is that the report gets released in the middle of the presidential campaign and then it becomes a political football," said a spokesman for Hastert.[12]

At that point, the Jersey Girls and the other FSC members realized that they had to draw a line in the sand.

"We had to send a message," said Beverly Eckert, "that this was an issue that transcended politics and that this Commission was on life support. If some kind of extension of time and budget wasn't granted, the entire investigation would have been compromised."

So, according to Lorie van Auken, they decided on a two-pronged approach. "First, we had never held a press conference at Ground Zero," she said. "Very few of us could even bring ourselves to going anywhere close to where our husbands had died. But we decided that we would send a trainload of people down to sit out in front of Hastert's office, and at the same time, for those who couldn't make that trip, we would hold a vigil where those Towers had stood."

In addition to the FSC's protest, Sen. John McCain threatened to delay a major public works bill if the extension wasn't granted. Within hours, on February 19, just a day before the widows were scheduled to hit Washington and Ground Zero, Hastert folded.

"The Speaker still thinks we need a report sooner rather than later," said a spokesman, "But he is willing to work with the president on this."[13]

Though the FSC wanted a year, the White House compromised at two months. The Commission would have until July 27 to finish its work.

It was another victory for the widows, but there were even more acute challenges ahead. First, while National Security Advisor Condoleezza Rice had met informally with some Commissioner staffers at the White Houses on a Saturday in early February, she refused to allow her "testimony" to be recorded or submitted under oath.[14]

Worse, President Bush had refused to commit on whether he would talk to the panel, and Commission staffers were complaining to Michael Isikoff of *Newsweek* about "maddening restrictions by White House lawyers on their access to key documents, particularly the August 6, 2001 PDB."[15]

Two days later, Isikoff and *Newsweek*'s Mark Hosenball made news again when they reported that, while researching his best-selling book *Bush at War, Washington Post* editor Bob Woodward had been given full access to many of the classified documents the White House was withholding from the 9/11 panel.[16]

"In his book Woodward claims he saw the PDBs," one Commission staffer said. "That was an argument that these were not the 'Holy of Holies."

"Clearly the White House had a double standard on this material," says Kristen Breitweiser. "It was becoming increasingly clear to us that, extension or not, if this panel didn't get full access to documents and witnesses and put them under oath in open session, then the American public would be cheated. And by the end of March that's how it was beginning to look."

Richard Clarke: Against New Enemies

Then something happened that kicked the 9/11 Commission into high gear. Suddenly, the hearings became front page news, earning gavel to gavel coverage on CNN. This new lease on life came not from Congress, or the White House, or even the intense lobbying of the Family Steering Committee.

On Sunday night, March, 21, 2004, Richard Clarke, the Bush Administration's former counterterrorism czar, appeared on *60 Minutes* and said that the President had "done a terrible job on the war against terrorism."[17] Clark told Lesley Stahl that on January 24, 2001, he had written a memo to Condoleezza Rice asking for an "urgent" cabinet level meeting "to deal with the impending al Qaeda attack. And that urgent memo wasn't acted upon."

Clarke went on to "blame the entire Bush leadership for continuing to work on Cold War issues when they came back in power in 2001." According to Clarke, in one conversation he warned Deputy Defense Secretary Paul Wolfowitz, "We have to deal with bin Laden, we have to deal with al Qaeda," to which the neo-conservative Wolfowitz replied, "No, no, no. We don't have to deal with al Qaeda. . . . We have to talk about Iraqi terrorism against the United States."

But Clarke insisted that he'd repeatedly told the president that the Saddam-bin Laden axis didn't exist. "I said, Mr. President, we've been looking at this. We've looked at it with an open mind and there's no connection."

The timing couldn't have been more fortuitous for those who hoped to light a fire under the 9/11 Commission. Four days after Clarke's memoir *Against All Enemies* was published, to an accompanying media blitz, he was due to testify before the Commission.

"Richard Clarke's testimony gave us hope," said Lorie van Auken. "Not only did he apologize at the hearing on March 24, which no official had ever done before, but he gave credibility to what many people had suspected for months—that the multiple failures in the intelligence community seemed to allow 9/11 to happen."

After Clarke's appearance, the floodgates opened. Despite White House opposition, Condoleezza Rice was forced to testify under oath. The controversial August 6 PDB was released. Then the Commission, which only three months earlier had contemplated cutting back hearings, heard twice from CIA Director George Tenet, along with the secretaries of state, FBI directors and joint chiefs of staff from both the Clinton and Bush administrations. The White House finally gave up on its restrictions to the Clinton documents, and the president himself submitted to questioning by all ten Commissioners in an informal meeting with Vice President Cheney on April 29.[18]

But in a signal of just how seriously some of the Commissioners were taking the investigation, two Democrats left the three and one-half hour session early, leading to a banner *New York Post* headline: "Oval Office Insult: Two 9/11 Dems Walk Out on Bush."[19]

"We were shocked," said Monica. "After all the months we had pushed the White House for full disclosure and access, now two of the *Democrats* were sending this signal, like they didn't seem to care. If the Pope had been in Washington that day and invited Kerrey and Hamilton to a special audience, they still should have stayed in front of the president and vice president questioning them."

"That walkout was a watershed event for us," said Patty Casazza. "When it came to 9/11 we were starting to see that this Commission was ignoring key evidence embarrassing to the government, no matter *what* political party you belonged to. The word 'cover-up' doesn't even seem to do it justice."

13

YEAR ONE DOGS AND PONIES

By mid-April, the *New York Times* was reporting that "With new evidence made public almost daily to show how the Sept. 11 attacks might have been prevented, the independent commission investigating them says its final report will offer a book-length chronology of the law enforcement intelligence and military failures that stopped the government from understanding the threat of Al Qaeda until it was too late."[1]

In the weeks following Richard Clarke's testimony, the panel had made strides in convincing the public that the Commission staff was leaving few stones unturned. Another *Times* piece, looking forward to the New York hearings in May, noted that "According to the Commission, some two dozen of its 80 staff members have read tens of thousands of documents, listened to hundreds of hours of audio tapes, including the 9/11 calls, and have conducted interviews with as many as 200 police, firefighters, civilians and Port Authority officials."

Earlier the *Times* had praised the Commission's "quiet but aggressive staff,"[2] which would eventually release seventeen Staff Statements summarizing the Commission's findings. By late May, staffers were quoted as saying that the final report would be built "on a framework provided by the interim staff reports . . . that have been mostly praised by members of the Commission."[3]

This book went to press shortly after the Commission's final report was published on July 22, 2004. Although the report contained more than one hundred pages of end notes, the body of the report closely reflected the Commission's staff statements which we'll analyze in depth in the pages that follow.

The end notes in the Commission report contained a number of new and intriguing allegations, but many of them referenced documents that were unavailable to the public—including the interrogation reports of Khalid Shaikh Mohammed, which were cited throughout the notes as the authoritative source for the date when the attacks were first planned. With those interrogation reports still sealed, any reader of the Commission's book-length report would be forced to take the staff's judgment of such sources on faith. And, as we'll see in our analysis of the Staff Statements to follow, the staff's accuracy and depth of research remains subject to serious question.

In fact, the staff didn't even begin to report its findings until more than a year after the investigation had commenced.

The First Year

The Commission's first four hearings revealed little or nothing that was new to the public body of knowledge on the intelligence failures that had led to the attacks. Several of the hearings even duplicated future proceedings.

"They seemed to be just marking time in those early days," says Mindy Kleinberg, who testified at the first hearing in the U.S. Customs House in New York on March 31, 2003. Lorie van Auken was more blunt. "It was kind of a dog and pony show—to let the media know, after four months, that the Commission was actually in existence."

According to the Commission's web site,* "The purpose of this public hearing was to engage those whose lives were forever changed by the

* http://www.9-11commission.gov.

events of September 11 in a public dialogue about the Commission's goals and priorities." The hearing opened with remarks from Governor George Pataki, Mayor Michael Bloomberg, and a wide ranging list of witnesses, from representatives of the Arlington, Virginia, fire department to a series of lawyers and scholars who praised the performance of the Bush administration in the war on terror.

Brian Jenkins, of the RAND Corporation, declared that "in the nineteen months since September 11, 2001, we have made considerable progress in destroying al Qaeda's base in Afghanistan and in disrupting its operational capabilities."[4]

Attorney Lee S. Wolosky, called to testify about al Qaeda's financial capabilities, reported that "Thanks to the hard work of the Bush administration over the past eighteen months, al-Qaeda's financial network has certainly been disrupted."[5]

On May 22, 2003, the first half of hearing Number Two proved to be a showcase for politicians. Under the subject heading of "congressional oversight," the panel first heard from Rep. Nancy Pelosi (D-CA), Sen. John McCain (R-AZ), and Sen. Joseph Lieberman (D-CT) about what they already knew—how the Commission came into being.

Then, four representatives of the Congressional Joint Inquiry* testified about their 800+ page report, which remained classified 153 days after it was filed.[6] None of them discussed exactly why Congress, which performs an oversight function on the intelligence community, had failed to detect a hint of the 9/11 attacks in the years since 1991, when Osama bin Laden first established a beachhead in New York City.

One extraordinary bit of "oversight" that was never mentioned in the declassified Joint Inquiry Report or the 9/11 Commission staff statements was a December 5, 1995, confidential memo to Sen. Orrin Hatch (R-UT), chairman of the Senate Judiciary Committee, from his staff. Titled "Investigation Into Terrorism," the memo proposed that the Committee investigate "the FBI's involvement before, during and after a terrorist threat and/or attack." The Judiciary Committee staff reported that "We have

* Former Sen. Bob Graham (D-FL), Sen. Richard Shelby (R-AL), Rep. Porter Goss (R-FL), and Rep. Jane Harmon (D-CA).

information that some instances, like the World Trade Center [bombing in 1993], could have been prevented."

This underscored evidence in the 1995 Day of Terror trial, when tapes were played quoting Nancy Floyd's undercover asset, Emad Salem, who believed that the FBI could have stopped Yousef before the 1993 Trade Center bombing.[7]

Knowing what we now know about the al Qaeda cell tied to Ramzi Yousef in 1996, that Senate investigation, if pursued at the time, would have given the Committee staff a much better sense of the al Qaeda threat. But that memo was never acted upon. No hearings were held on the staff's recommendations, and no investigative body examining congressional oversight has mentioned it since.[8] And there wasn't a word about it in the 9/11 Commission report, which was critical of Congress for failing to properly oversee the intelligence community.

Entertaining the Yousef-As-Iraqi-Agent Theory

The third hearing on July 8, 2003 offered a showcase for one of the Bush Administration's chief apologists for the invasion of Iraq. Laurie Mylroie, a scholar with the neo-conservative American Enterprise Institute (AEI), had for years been the chief proponent of the theory that Ramzi Yousef was actually an Iraqi agent who assumed the identity of a Kuwaiti named Abdul Basit Karim.[9] Her primary allies in proposing the theory were Paul Wolfowitz, now the deputy secretary of defense, and his then-wife Clare, who was credited by Mylroie with having "fundamentally shaped the book."

Mylroie also enjoyed the support of former CIA Director R. James Woolsey and James Fox, the deceased former ADIC of the FBI's New York office, on whose watch Yousef built the 1993 WTC bomb. Fox was later dismissed by FBI Director Louis Freeh, for denying publicly and angrily that the Bureau could have prevented the attack.[10]

In the December 2003 issue of *Washington Monthly,* respected terrorism expert Peter Bergen[11] wrote a devastating portrait of Mylroie,[12] who he noted had been "an apologist for Saddam's regime, but reversed her position upon his invasion of Kuwait in 1990 and with the zeal of the academic spurned, became rabidly anti-Saddam."

In the piece, subtitled "Laurie Mylroie: The Neocons' Favorite Conspiracy Theorist," Bergen noted that until the Gulf War, Mylroie had been a respected scholar who had taught at both Harvard and the U.S. Naval War College, and co-authored with *New York Times* reporter Judith Miller a book called *Saddam Hussein and the Crisis in the Gulf,* a "well-reviewed bestseller translated into more than a dozen languages."

For the rest of the piece, Bergen carefully recited the best intelligence that is known about Ramzi Yousef and his ties to Osama bin Laden and al Qaeda—not Iraq.

He began by citing the testimony of al Qaeda turncoat Jamal al-Fadl, who told a New York jury in 2000 that he had seen Yousef in the al Qaeda Sadda training camp on the Pakistan-Afghan border between 1989 and 1991. Next, Bergen noted that Yousef's Bojinka co-conspirator Wali Khan Amin Shah was an intimate of Osama bin Laden's, and that Yousef's uncle Khalid Shaikh Mohammed had not only sent money to his nephew for the 1993 WTC attack but would go on to become "al Qaeda's military commander and the chief planner of 9/11."

Next Bergen quoted Vince Cannistraro, who had run the CIA's Counterterrorism Center in the early 1990s. "My view," said Cannistraro, "is that Laurie has an obsession with Iraq and trying to link Saddam to global terrorism. Years of strenuous effort to prove the case have been unavailing." Bergen even quoted Kenneth Pollack, the author of *The Threatening Storm: The Case for Invading Iraq,* who also dismissed Mylroie's theories. "The NSC had the intelligence community look very hard at those allegations that the Iraqis were behind the 1993 World Trade Center attack," said Pollack, a former CIA analyst. "Finding those links would have been very beneficial to the U.S. government at the time, but the intelligence community said that there were no such links."

"The point," wrote Bergen, "is that the 1993 attack was plotted not by Iraqi intelligence, but by men who were linked to al Qaeda." And he went on to make the case that al Qaeda is dominated not by Iraqis, but by Egyptian radicals, including Dr. Ayman al Zawahiri and Sheikh Omar Abdel Rahman.

But the 9/11 Commissioners and staff allowed Mylroie to take up a substantial part of that third hearing, expounding a specious theory that supported the Bush Administration's Iraqi invasion. In the end she concluded that "the major terrorist strikes against the U.S. that were attributed to 'loose networks' of Islamic militants, including al Qaeda, are much better explained as Iraq, working with and hiding behind the militants."[13]

The Buck Stops with the President and the CIA Director

The fourth hearing, held in Washington on October 14, was the last for Max Cleland, who began the questioning in the afternoon session by quoting a passage from *1000 Years for Revenge.* His first question was directed at John Gannon, the staff director of the House Select Committee on Homeland Security.

"The most recent book on 9/11, called *1000 Years for Revenge* . . . says, 'the mass murders of September 11 were the culmination of a plot that had been in the works for seven years,'" said Cleland, reading from the book. "The pathology that spawned it had begun to infect our country another five years before that. Twelve years. That's how far back I followed the trail to 9/11. With all the dots on that line, why couldn't our government see it coming?' My question . . . starting with Mr. Gannon . . . who is responsible for warning this country of an attack on this nation and who's accountable?"[14]

"I believe the president of the United States is responsible," Gannon responded, "but I also believe that his major intelligence advisor should be and must be the Director of Central Intelligence as a single focus on the warning."

Earlier that day, the Commission had heard from two former CIA directors, James M. Schlesinger, who served under Gerald Ford, and John Deutch, who served from May 1995, when the Philippines National Police first learned of Ramzi Yousef's involvement in the 9/11 plot, through December 1996, six months after the crash of TWA 800.

While Deutch in his testimony recommended a sweeping reform of the U.S. intelligence community,[15] none of the Commissioners sought to ask him about his own knowledge of the 9/11 plotters at the time of his tenure.

Monica Gabrielle, who attended the hearing, reflected the growing sense of skepticism at that time among the widows. "If, as Gannon, said, the buck should stop with the president and the CIA director, I was sitting there wondering why the Commissioners didn't ask Deutch about *his* role in the intelligence failures that led to the attack."

As mentioned, the next hearing, held on November 19 at Governor Kean's home venue, Drew University, was virtually ignored by the media. Except for statements by Monica Gabrielle and Sally Regenhard, the hearing seemed to drift far beyond any attempt by the Commission to fulfill its mandate and examine the causes of the 9/11 attacks.

The testimony included a lengthy statement by John Degnan, vice chairman of the Chubb Group, who noted that his company was "one of the largest insurers of the World Trade Center and its tenants, and one of the largest insurers of businesses, homes and automobiles in Lower Manhattan as well."

Degnan then launched into a protracted homage to the Chubb Group, which he noted had dispatched a "swat team" of claims adjusters to Ground Zero when air travel was grounded and movement in and around Lower Manhattan severely restricted. "Chubb senior management," Degnan testified, "then set about perhaps its most important task: a detailed analysis of whether or not the 'war exclusion,' common language in all U.S. insurance policies, would apply; in which case the monumental property losses would not be covered."

Complimenting his company, Degnan then declared proudly that "We balanced the legal, moral and business ramifications of such an analysis, and determined that the war exclusion did *not* apply—a decision that allowed Chubb to publicly state our coverage position ahead of any other insurer. We believe that in doing so we influenced the conclusion later reached by the rest of the insurance industry."

"Putting aside whether it was appropriate to allow the Commission's time to be taken up by what amounted to a commercial for one company," says Lorie van Auken, "you have to ask, what company *ever* could have gotten away with denying insurance claims at that point? But none of this mattered to us. We were sitting there at that hearing with pits in our stomachs, wondering whether this Commission which we had fought so hard to establish, was ever going to find out why our husbands had died."

It took another hearing, which included a lengthy analysis of the issue of "preventive detention" and the use of enemy combatant laws to combat terrorism, before the Commission finally went into high gear.

"As far as we were concerned," says Patty Casazza, "that entire first year—two thirds of this Commission's existence—was a virtual waste."[16]

14

WARNING: PLANES AS WEAPONS

The first of the Commission's seventeen staff reports was issued at the seventh hearing, conducted over two days from January 26–27, 2004. Described by one *New York Times* reporter as "gripping" and "a good read,"[1] the reports, which rolled out over six hearings in the space of six months, reached some dramatic conclusions.

The Commission put to rest, at least for a time,[2] false allegations made by the Bush Administration before the Iraq invasion that associated Saddam Hussein with Osama bin Laden in perpetrating the 9/11 attacks.[3] Further, in perhaps the most sweeping series of conclusions during the twelfth hearing, in mid-June, the Commission staff described widespread confusion in the Bush administration on the day of 9/11, which at times bordered on chaos.[4]

As stunning as that pronouncement was, and as gripping as the Commission staff's minute-by-minute account was, entire areas of Staff Statement No. 17 conflict with what has been in the public record for years. In the pages ahead we'll devote an entire chapter to an analysis of the Commission's findings on the day of 9/11 itself, measuring the staff's conclusions against thousands of open source news reports compiled by Paul Thompson and the Center for Cooperative Research.

Taking the "Mastermind's" Word

The staff statements also gave the Saudis what amounted to a free pass on 9/11,[5] even though fifteen of the nineteen hijackers were from Saudi Arabia, and compelling evidence from the Philippines National Police (PNP) indicates that Osama bin Laden's brother-in-law, the Saudi Mohammed Jamal Khalifa, funded the Manila Yousef-KSM cell that spawned the plots.[6]

As noted earlier, the Commission's conclusion that Khalid Shaikh Mohammed began planning for the plot on his own in 1996 is based on Mr. Mohammed's own reported allegations, and conflicts with a depth of intelligence furnished to the Justice Department by the PNP in 1995. That material puts the genesis of the plot in 1994, and Ramzi Yousef at its heart.

In this chapter, we'll summarize the Commission's most significant findings and offer critical analysis of how far the staff went in uncovering the full truth. As the reader will see, the quality of the staff statements is uneven. Some provide a no-holds-barred examination of lapses in intelligence and defense. Others offer a watered-down version of events, eliminating significant facts that were uncovered by the Joint Inquiry's investigation or stated repeatedly in the media. But perhaps most significant in the eyes of the victims' families, the statements fail to assess blame, or reach conclusions of fault.

Staff Statement No. 1: Entry of the 9/11 Hijackers into the United States

The most startling finding in this report is that Khalid Sheikh Mohammed, who was indicted in 1996, got a legal visa to enter the United States on July 23, 2001, about six weeks prior to 9/11. "Although he is not a Saudi citizen," said the staff, "and we do not believe he was in Saudi Arabia at the time, he applied for a visa using a Saudi passport and an alias, Abdulrahman al Ghamdi." The visa was granted under the program known as Visa Express, designed to expedite the entry of Saudis into America. A number of the 9/11 hijackers took advantage of the program: Beginning in 1997 they submitted a total of twenty-four applications, and received twenty-four visas.[7]

Staff Statement No. 2: Three 9/11 Hijackers: Identification, Watchlisting, and Tracking

In this report the Commission staff goes particularly easy on the CIA when describing the extraordinary agency blunder in failing to fully report the surveillance of hijackers Khalid al-Midhar and Nawaf al-Hazmi at the January 2000 9/11 planning session in Kuala Lumpur, Malaysia. In fact, the description of the CIA performance is almost flattering. On getting advance notice of the meeting from the National Security Agency (NSA), the staff declares, "both CIA Headquarters and U.S. officials around the world began springing energetically into action."[8]

This meeting would take on enormous significance later, when the CIA learned that one al Qaeda leader at the meeting was Khallad bin Atash, who ten months later would participate in the bombing of the U.S.S. *Cole,* which killed seventeen Americans.

"CIA Headquarters asked NSA to put al Midhar on that agency's watch list, which had limited effectiveness," the report notes. "There was no other effort to consider the onward destinations of these Arabs and set up other opportunities to spot them."

Even a year later, when the CIA had learned that bin Atash had attended the meeting in Kuala Lumpur, nobody at the agency took the initiative to pick up the trail of al-Midhar and al-Hazmi, who were soon living openly in San Diego under their own names. The failure of the CIA and the FBI to connect the dots on two of the 9/11 hijackers who were on the agency's radar twenty-one months before 9/11, was considered by the Joint Inquiry to be one of the most significant intelligence failures on the road to September 11. But the 9/11 Commission staff statement lets this incredible blunder go by in passing, without a rebuke to the CIA.

Later in the statement, the staff notes that "The director of the Counterterrorism Center at the time, Cofer Black, recalled to us that this operation was one among many and that, at the time, 'it was considered interesting, but not heavy water yet.'" Then, instead of chastising Black and the onetime head of his al Qaeda unit, the staff singles them out for praise, blaming "the system" rather then the men:

"We believe both Mr. Black and the former al Qaeda unit head are capable veterans of the Directorate of Operations, among the best the Agency has produced. Therefore we find these accounts more telling about the system than about the people. In this system no one was managing the effort to insure seamless handoffs of information or develop an overall interagency strategy for the operation."[9]

Black, who is now counterterrorism coordinator for the State Department, predicted in January 2004 that 70 percent of the al Qaeda network had been neutralized. "They are being hunted down," asserted Black; "their days are numbered."[10] His statement came two months before al Qaeda was connected to the Madrid train bombing, which killed 191 people and injured more than 1800.

Staff Statement No. 3: The Aviation Security System and the 9/11 Attacks

This staff statement is fraught with lapses and contradictions. It begins with the assertion that: "No U.S. flagged aircraft had been bombed or hijacked in over a decade. Domestic hijacking in particular seemed like a thing of the past, something that could only happen to foreign airlines that were less well protected."

In fairness to the Commission staff, it did not yet have the new information on TWA 800 at the time the statement was issued. By early April, however, Angela Clemente and Dr. Stephen Dresch had presented their evidence to the Commission, citing the Scarpa-Yousef kites and FBI 302s; yet the staff made no mention of this information in later statements.

Then, on page 3, the staff contradicts its own later finding from Staff Statement No. 16 that Khalid Shaikh Mohammed began planning the 9/11 attacks in 1996.

"Some time during the late 1990's," the statement says, "the al Qaeda leadership made the decision to hijack large, commercial multi-engine aircraft and use them as devastating weapons, as opposed to hijacking a commercial aircraft for use as a bargaining tool."

This statement not only ignores the evidence uncovered by Col. Mendoza of the PNP that Yousef and KSM hatched the plot in 1994, but runs counter to the findings of the Joint Inquiry, upon which the 9/11 Commission was expected to build. Below is a verbatim excerpt from the Joint Inquiry's final declassified report, which describes in detail a dozen separate plots from 1994 to 2001 in which Islamic terrorists considered using aircraft as weapons:

> INTELLIGENCE INFORMATION ON POSSIBLE TERRORIST USE OF AIRPLANES AS WEAPONS[11]
>
> The Joint Inquiry confirmed that, before September 11, the Intelligence Community produced at least twelve reports over a seven-year period suggesting that terrorists might use airplanes as weapons. As with the intelligence reports indicating Bin Laden's intentions to strike inside the United States, the credibility of sources was sometimes questionable and information often sketchy. The reports reviewed by the Joint Inquiry included:
>
> - In December 1994, Algerian Armed Islamic Group terrorists hijacked an Air France flight in Algiers and threatened to crash it into the Eiffel Tower. French authorities deceived the terrorists into thinking the plane did not have enough fuel to reach Paris and diverted it to Marseilles. A French anti-terrorist force stormed the plane and killed all four terrorists.
> - In January 1995, a Philippine National Police raid turned up material in a Manila apartment suggesting that Ramzi Yousef, Abdul Murad, and Khalid Shaykh Mohammad planned, among other things, to crash an airplane into CIA Headquarters.* The police said that the same group was responsible for the

*As reported in *1000 Years for Revenge,* this was an early admission by Ramzi Yousef's partner Abdul Hakim Murad to Col. Rodolfo Mendoza, his interrogator from the Philippines National Police. As the 67 days of his questioning wore on, Murad admitted that the plot was much more elaborate involving 6 targets on both U.S. coasts. Murad confessed that Islamic pilots were then training for the attacks in U.S. flight schools. After Yousef's capture in February 1995, his uncle Khalid Shaikh Mohammed executed the plot, which culminated on Sept. 11, 2001.

bombing of a Philippine airliner on December 12, 1994. Information on the threat was passed to the FAA, which briefed U.S. and major foreign carriers.

- In January 1996, the Intelligence Community obtained information concerning a planned suicide attack by persons associated with Shaykh al-Rahman and a key al-Qa'ida operative to fly to the United States from Afghanistan and attack the White House.

- In October 1996, the Intelligence Community obtained information regarding an Iranian plot to hijack a Japanese plane over Israel and crash it into Tel Aviv. A passenger would board the plane in the Far East, commandeer the aircraft, order it to fly over Tel Aviv, and crash the plane into the city.

- In 1997, an FBI Headquarters unit became concerned about the possibility that an unmanned aerial vehicle (UAV) would be used in terrorist attacks. The FBI and CIA became aware of reports that a group had purchased a UAV and concluded that the group might use the plane for reconnaissance or attack. The possibility of an attack outside the United States was thought to be more likely, for example, by flying a UAV into a U.S. embassy or a U.S. delegation.

- In August 1998, the Intelligence Community obtained information that a group, since linked to al-Qa'ida, planned to fly an explosive-laden plane from a foreign country into the World Trade Center. As explained earlier, the FAA found the plot to be highly unlikely given the state of the foreign country's aviation program. Moreover, the agencies concluded that a flight originating outside the United States would be detected before it reached its target. The FBI's New York office took no action on the information.

- In September 1998, the Intelligence Community obtained information that Bin Laden's next operation might involve flying an explosives-laden aircraft into a U.S. airport and detonating it. This information was provided to senior government officials in late 1998.

- In November 1998, the Intelligence Community obtained information that the Turkish Kaplancilar, an Islamic extremist group, had planned a suicide attack to coincide with celebrations marking the death of Ataturk, the founder of modern Turkey. The conspirators, who were arrested, planned to crash an airplane packed with explosives into Ataturk's tomb during a ceremony. The Turkish press said the group had cooperated with Bin Laden, and the FBI's New York office included this incident in a Bin Laden database.

- In February 1999, the Intelligence Community obtained information that Iraq had formed a suicide pilot unit that it planned to use against British and U.S. forces in the Persian Gulf. The CIA commented that this was highly unlikely and probably disinformation.

- In March 1999, the Intelligence Community obtained information regarding plans by an al-Qa'ida member, who was a U.S. citizen, to fly a hang glider into the Egyptian Presidential Palace and detonate explosives. The person, who received hang glider training in the United States, brought a hang glider to Afghanistan. However, various problems arose during the testing of the glider. He was subsequently arrested and is in custody abroad.
- In April 2000, the Intelligence Community obtained information regarding an alleged Bin Laden plot to hijack a Boeing 747. The source, a "walk-in" to the FBI's Newark office, claimed that he had learned hijacking techniques and received arms training in a Pakistani camp. He also claimed that he was to meet five or six persons in the United States. Some of these persons would be pilots who had been instructed to take over a plane, fly to Afghanistan, or, if they could not make it there, blow the plane up. Although the source passed a polygraph, the Bureau was unable to verify any aspect of his story or identify his contacts in the United States.*
- In August 2001, the Intelligence Community obtained information about a plot to bomb the U.S. embassy in Nairobi from an airplane or crash the airplane into it. The Intelligence Community learned that two people who were reportedly acting on instructions from Bin Laden met in October 2000 to discuss this plot.

Staff Statement No. 4: The Four Flights

This statement offers a brief analysis of the security breaches that led to the 9/11 hijackers boarding the two Boeing 767s that hit the World Trade Center, American Airlines Flight 11 and United Airlines Flight 175, as well as the two hijacked 757s: AA Flight 77, which hit the Pentagon, and UA Flight 93, the plane that crashed in Pennsylvania. The staff statement accepts the widely held belief that Flight 93 was brought down when a group of passengers stormed the cockpit; this would be contradicted by the Commission's final report, which presented transcripts from Flight 93's cockpit recorder, and suggested that the passengers never, in fact, made it into the cockpit, and that the hijackers downed the plane themselves. Neither account focuses on the conflicting evidence suggesting

*This story of Niaz Khan, a U.K. citizen of Pakistani origin, will be discussed in Chapter 17.

that the plane may have been shot down. We'll examine that issue in more depth in Chapter 20.

In Staff Statement No. 4, the Commission staff explains how the relatively untrained hijackers could have done such precision flying that day, noting that the Flight Management Computer in each aircraft "could be programmed in such a manner that it would navigate the aircraft automatically to a location of the hijackers choosing . . . at a speed and altitude they desired, provided the hijackers possessed the precise positioning data necessary."[12]

The staff report concludes by noting that "The witnesses' accounts of the phone calls [from the hijacked planes] are consistent and are quite specific about the kind of weapons that were reported present—knives, mace, and a bomb—as well as the nature of the assaults on board—the 'stabbing' of two crew members and a passenger."

The Commission explains the presence of knives on board by noting that while FAA regulations at the time forbade passengers from bringing box cutters aboard, they were permitted to carry knives with blades up to four inches long.

But the staff doesn't explain how mace or a bomb, both clearly forbidden by FAA regulations, passed airport security. That question becomes important in light of a report in the *Washington Post* on March, 2, 2002 that nine of the 19 hijackers were selected for "special security screenings" before they boarded their flights, including two who were singled out because of irregularities in their IDs.[13]

Staff Statement No. 5: Diplomacy

This staff statement contains some of the most significant misrepresentations and half-truths in all of the Commissions' findings. In fact, the report describes a certain level of efficiency by U.S. intelligence agencies that belies the real depth of their failures leading up to 9/11. Case in point. The Commission's description of the capture of Ramzi Yousef and the search for his uncle Khalid Shaikh Mohammed. We'll contrast the verbatim conclusions of the staff statements with the facts as they have been uncovered by various reporters, lawyers, and investigators in the years since the first World Trade Center bombing in 1993. All of this information, including the findings set forth in multiple al Qaeda related federal trials, was on the public record and available to the Commission staff.

Ramzi Yousef

The 9/11 Commission: "The U.S. government believed that the World Trade Center attack of 1993 had been carried out by a 'cell' led by Abdul Basit Mahmud Abdul Karim, better known by his alias, Ramzi Yousef. Yousef had escaped and was a fugitive. By early 1995 he was also wanted for his participation in a plot to plant bombs on a dozen American airliners in the Far East. Yousef had fled to Pakistan. The United States learned where he was and, working effectively with Pakistani officials, carried out a rendition that sent him back to America for trial."[14]

The facts: Yousef was apprehended due to some extraordinary luck and two significant factors: the accidental fire in Room 603 of the Dona Josefa apartments, which caused him to flee; and the confession of Istaique Parker, an Islamic South African "walk-in" to the U.S. Embassy in Islamabad, who decided to betray Yousef in return for a $2 million reward sponsored by the U.S. State Department.

Rather than working "effectively" with Pakistani officials, sworn testimony in the Bojinka cases suggests another scenario. According to the testimony, Brad Garrett, a Washington-based FBI agent who had arrived in Islamabad the morning of the capture (looking for another suspect),[15] was late arriving at the Su Casa guesthouse, where agents of the DEA and Diplomatic Security Services (working with the Pakistani ISI) had already removed Yousef.[16]

Then, apparently, Garrett failed to canvas the small guesthouse, which turned out to have been controlled by a nongovernmental organization (NGO) funded by Osama bin Laden. If he had done a routine canvas, Garrett might have come upon Yousef's uncle Khalid Shaikh Mohammed, who was staying in a downstairs room. In an audacious display of bravado, KSM actually gave an account of the Yousef takedown to a reporter for

Time magazine, using his own name, "Khalid Shaikh."[17] Before Garrett could get there, however, the "mastermind of 9/11" escaped. Later, in a *60 Minutes II* story, Garrett took personal credit for Yousef's capture.[18]

Khalid Shaikh Mohammed

The 9/11 Commission: In 1995, the United States also learned that Khalid Sheikh Mohammed, "KSM," was living in Doha, Qatar, and was reportedly employed by a government agency there. The United States obtained other specific details that could have helped locate KSM, who was then sought as a suspect in the Ramzi Yousef airlines plot. Working with the U.S. ambassador in Doha, the FBI and CIA worked on how to capture KSM. But they were reluctant to seek help from the Qatari government, fearing that KSM might be tipped off. The U.S. government instead considered the option of capturing KSM without Qatari help. The available options were rejected as unwieldy and too risky. Therefore, after first waiting for a sealed indictment against KSM to be handed down by a New York grand jury, in January 1996 the U.S. government asked the Emir of Qatar for help.

Qatari authorities first reported that KSM was under surveillance. They then asked for development of an alternative plan that would conceal their aid to Americans. They then reported that KSM had disappeared. KSM would later become a principal planner of the 9/11 attacks, and was captured in 2003. We do not know whether KSM was tipped off in 1996.

The facts: What the Staff Statement doesn't say is that while hiding out in Qatar, a major U.S. ally in the Persian Gulf, KSM was believed to be staying under the protection of one Abdullah bin Khalid al-Thani, the interior minister of Qatar and a member of its ruling family. Al-thani was also the nation's security chief. As first reported by Bill Gertz in the *Washington Times,*[19] the story was later advanced by reporters for ABC News[20] and the *Los Angeles Times.*[21] According to their investigations, this is what actually happened:

After former CIA case officer Bob Baer tipped the FBI to KSM's presence in the Qatari capital, the elite FBI Hostage Rescue Team was dispatched to Doha. But Qatari officials reportedly told the Feds that they wanted to "put the handcuffs" on Mohammed themselves. So while the U.S. agents cooled their heels waiting, al-Thani reportedly helped KSM to escape. He was spirited off to the Czech Republic with new fake ID in the name of Mustafa Nasir.[22] Informed of this alleged betrayal by al-Thani, a senior official of a friendly government, Richard Clarke expressed shock. "You're telling me that [al-Thani] is today in charge of security inside Qatar? I hope that's not true," said Clarke.[23]

Consider that Doha, Qatar, was the location of CENTCOM, the Pentagon's central command for the invasion of Iraq in March 2003. One of the Bush Administration's primary justifications for the preemptive strike on Baghdad was Saddam Hussein's alleged connection to al Qaeda and involvement in the 9/11 plot, which have since been disproved and denounced by the 9/11 Commission. But since these news reports in 2002 there has been evidence on the public record that a top ranking Qatari official may have been a co-conspirator in helping the 9/11 mastermind to flee. Still, the Commission makes no mention of al Thani or the incident in any of its staff statements, and only a glancing reference to the Qatari minister in its final report.

It is unknown whether Defense Secretary Donald Rumsfeld was aware of the al-Thani/KSM ties at the time the United States signed a cooperation agreement with Qatar to base its command there. But since KSM was then not only acting to perfect the 9/11 plot, but communicating with his nephew Ramzi, who was preparing to plant a bomb aboard a plane, the implications of Mohammed's escape are immense.

"This is a contradiction that boggles the mind, when you think about it," says Lorie van Auken. "I mean, what can the 9/11 Commission's findings be worth if they miss that kind of a lapse and leave it out of the record?"

As we'll see in the chapters to follow, they missed so much more.

15

"ALARMING THREATS" POURING IN

Overall, the four staff statements that rolled out over a two-day period during the March 23–24, 2004, hearings were among the most candid and unvarnished the Commission produced. While equivocating to some degree, they leave the reader with the impression that if the Commission had had more time, better funding, and more objective staff members—and had they been less influenced by election-year politics—the final report of the Commission might have offered a more honest rendering of the truth.

Staff Statement No. 6: The Military

This statement, one of the most powerful generated by the Commission, describes an intelligence apparatus at the highest levels of the Clinton and Bush Administration that seemed impotent to attack the al Qaeda threat, despite mounting evidence that Osama bin Laden was prepared to inflict mass casualties on U.S. citizens.

The statement portrays Clinton's national security director, Sandy Berger, as lacking the resolve to take military action in Afghanistan where bin Laden was based after cruise missile strikes against him in the Sudan missed their mark.[1] Following those strikes in 1998, the staff report says that Richard Clarke recommended hitting al Qaeda training camps, but Joint Chiefs chairman Hugh Shelton balked. He was, according to the statement, reluctant to use a "very expensive missile to take out 'jungle gym'" training camps.

Fearing that further unsuccessful strikes would only increase "bin Laden's stature," Shelton was seconded by Deputy NSA advisor James Steinberg, who opined that such strikes offered "little benefit, [and] lots of blowback against [a] bomb-happy U.S." Shelton, the nation's top military man, was worried about another *Black Hawk Down*-style embarrassment, referencing the downing of two U.S. choppers in Somalia by al-Qaeda linked forces in 1993.

The report notes that DOD leaders continually complained about the lack of "actionable intelligence." But the statement lists three separate occasions when bin Laden was sighted, offering opportunities to kill or capture him: He was spotted at a desert camp in February 1999 and twice in Kandahar (in December 1998 and May 1999), but no action against him was taken. A CIA bin Laden unit chief reportedly complained that "having a chance to get OBL three times in 36 hours and foregoing the chance each time has made me a bit angry." But the bin Laden hunt didn't fare any better on President Bush's watch. In fact, by January 2001 there was even more compelling evidence of the Saudi billionaire's culpability, following the attack on the U.S.S. *Cole.*

According to the staff report, when Condoleezza Rice's deputy Stephen Hadley was presented with similar strike options against bin Laden five days into Bush's presidency, he rejected the option, describing such missile strikes as "tit for tat." As the staff tells it, Hadley said the Bush White House's response to the *Cole* attack would be "a new, more aggressive strategy." But no such strategy was ever implemented prior to 9/11. In fact, Paul Wolfowitz, the deputy secretary of defense and the prime architect of the Iraqi invasion, told the staff that by the time Bush had moved into the White House the Cole incident was "stale."

SECDEF Donald Rumsfeld was briefed on fifty items by his predecessor, William Cohen, during the transition. Included in the briefings was intelligence on "bin Laden and programs related to domestic preparedness against terrorist attacks and weapons of mass destruction." But Rumsfeld told the Commission staff that "he did not recall what was said about Bin Laden" at those briefings. More revealing, in terms of the Bush administration's defense priorities, was the testimony of Undersecretary for Policy Douglas Feith. He admitted that when he took office in July 2001 Rumsfeld told him to focus on "working with the Russians . . . to dissolve the Anti-Ballistic Missile (ABM) Treaty and preparing a new nuclear arms control pact." And Rumsfeld himself admitted to the panel staff that "He did not recall any particular counterterrorism issue that engaged his attention before 9/11."

And yet by late July 2001, as noted in the very next Staff Statement, the intelligence community was receiving "indications of multiple, possibly catastrophic terrorist attacks being planned against American interests overseas."[2]

Staff Statement No. 7: Intelligence Policy

This Staff Statement, which was also straightforward and unprotective of either administration, contained a revelation that explained why the CIA couldn't produce any "actionable intelligence" that might have led to the killing or capture of bin Laden. "[T]he United States had no diplomatic or intelligence officers living and working in Afghanistan," noted the report. With no HUMINT (human intelligence) on the ground, the United States was left to negotiate with the hard-line Taliban government, which was sheltering bin Laden in the first place. So the CIA was forced to develop a network of "proxies," local freelance "agents" who were paid to inform, and thus highly unreliable.

The Clinton administration had a military option: the Predator, a drone-like remotely piloted surveillance vehicle which was being prepped to carry missiles. By the fall of 2000 Richard Clarke was pushing the Pentagon to use it, but CIA director George Tenet, backed by the

Joint Chiefs, balked. The staff report noted candidly that Clarke believed that the arguments (against using the Predator) were "stalling tactics by the CIA's risk-averse Directorate of Operations."

By July 2001, Rice's deputy Stephen Hadley ordered that the reconnaissance aircraft be prepped for deployment by September 1. On September 11, however, the Commission staff was told by the Air Force program manager that they were still working out the aircraft's "technical issues."

To underscore the negligence of the Bush Administration in ignoring the al Qaeda threat in the summer of 2001, the staff statement relates the story of two veterans of the CIA's Counterterrorism Center (CTC) who were "deeply involved in OBL issues." The CTC members, according to the Commission report, were so worried about an impending disaster, "that one of them . . . considered resigning and going public with their concerns." But DCI Tenet insisted to the 9/11 Commission staff that he believed White House officials "grasped the sense of urgency he was communicating to them."

Then the staff statement, which began with such candor in its description of CIA failings, ends with another pat on the back for both Director Tenet and his staff: "The Commission has heard numerous accounts of the tireless activity of officers within the CTC and the OBL Station trying to tackle al Qaeda before 9/11. DCI Tenet was also clearly committed to fighting the terrorist threat."

Still, the next Statement, while pulling some punches, would document the most devastating intelligence lapses of all by the Bush White House in the months leading up to 9/11.

Staff Statement No. 8: National Policy Coordination

This report describes in detail the contrast between the Clinton and Bush Administration when it came to counterterrorism policy. Once again, Richard Clarke was the key.[3] Under Clinton, Clarke coordinated the work of the CSG (Counterterrorism And Security Group) and was part of what was known as "The Small Group," including principals such as the vice president, the DCI, the FBI director, the secretaries of state and defense, and the CJCS (Joint Chiefs Chairman).

But in January 2001, when Bush took office, Clarke was relegated by incoming NSA Condoleezza Rice to a position that obliged him to report to a "Deputies Committee" chaired by her number two, Stephen Hadley. While on operational matters Clarke was permitted to deal directly with Rice, he told the Commission staff that he considered the move a "demotion." It came at a time when the al Qaeda drums were beating louder and louder with each passing day. Consider this verbatim excerpt from Staff Statement No. 8:

> From April through July, alarming threat reports were pouring in. Clarke and the CSG were consumed with coordinating defensive reactions. In late June, Clarke wrote Rice that the threat reporting had reached a crescendo. Security was stepped up for the G-8 summit in Genoa, including air-defense measures. U.S. embassies were temporarily closed. Units of the Fifth Fleet were redeployed from usual locations in the Persian Gulf. Administration officials, including Vice President Cheney, Secretary Powell, and DCI Tenet, contacted foreign officials to urge them to take needed defensive steps.
>
> On July 2, the FBI issued a national threat advisory. Rice recalls asking Clarke on July 5 to bring additional law enforcement and domestic agencies into the CSG threat discussions. That afternoon, officials from a number of these agencies met at the White House, following up with alerts of their own, including FBI and FAA warnings. The next day, the CIA told CSG participants that al Qaeda members "believe the upcoming attack will be a 'spectacular,' qualitatively different from anything they have done to date."
>
> On July 27 Clarke reported to Rice and Hadley that the spike in intelligence indicating a near-term attack appeared to have ceased, but he urged them to keep readiness high; intelligence indicated that an attack had been postponed for a few months.

It was during this period that President Bush went off on his controversial thirty-day "working vacation" at his Crawford, Texas, ranch. Taking a month off after only six months in office, Bush seemed undeterred by the criticism at the time. In a *USA Today* poll released on August 6, the day he received the notorious Presidential Daily Briefing, 55 percent of those questioned felt that the President was taking too much time off.[4]

The Staff Statement paints a picture of an Administration whose various intelligence components seem disconnected from the impending al Qaeda threat. Describing the PDB, which at that point had not been declassified, the March 24, 2004, staff statement notes that "neither the White House nor the CSG received specific, credible information about any threatened attacks in the United States. Neither Clarke nor the CSG were informed about the August 2001 investigations that produced the discovery of suspected al Qaeda operatives in the United States. Nor did the group learn about the arrest or FBI investigation of Zacarias Moussaoui in Minnesota."

What the staff statement didn't say was that CIA Director Tenet had received a report on Moussaoui—the flight student with al Qaeda ties who had paid instructors to teach him how to take off in a jumbo jet, but didn't particularly care how to land.

While the President was vacationing in Crawford, a report was forwarded to the DCI entitled "Islamic terrorist learns to fly." A month earlier, on July 10, FBI Headquarters and the New York office had received the celebrated "Phoenix memo," suggesting that Islamic pilots should be monitored in U.S. flight schools. Tenet later told the 9/11 staff that at that point in the summer "the system was blinking red."[5] Then, on September 4, barely a week before 9/11, Tenet attended a principals meeting with key intelligence officials including Richard Clarke, but *incredibly, he failed to bring up the Moussaoui report.*[6]

The staff statement ended on a chilling note. Just before that September 4 principals meeting, Richard Clarke wrote to Rice "summarizing many of his frustrations." Like a counterterrorist Paul Revere, he "urged policymakers to imagine a day after a terrorist attack, with hundreds of Americans dead at home and aboard and ask themselves what they could have done earlier." The statement makes no mention of any reply from the National Security Advisor.

Showdown with Condoleezza Rice

At this point in late March 2004, the staff of the 9/11 Commission seemed to be throwing the gauntlet down to Rice. While she had agreed to meet informally with staffers for a four-hour session in February, Rice had resisted any suggestion that she appear before the Commission in open session and under oath.[7] Although she was clearly the linchpin of the Bush administration's counterterrorist policies in the months leading up to the attacks, she remained off-limits to the Commission, even as she was pressing her views in the media, from an op-ed page piece to a session with Ed Bradley on *60 Minutes.*[8] The day after that broadcast, the 9/11 Commission ratcheted up the confrontation with the White House. They were now *demanding* that Rice appear.[9]

In the midst of all this, Commissioner Jamie Gorelick, one of the two panelists with full access to all the secret documents, said that some Bush administration files from the summer of 2001 "would set your hair on fire."[10] Following Richard Clarke's testimony before the Commission on March 24, approval of the president's handling of homeland security and the war on terror had dropped by eight points.[11]

So, by March 30, in another stunning back-down reminiscent of the reversal on the Commission's own extension, the White House relented—setting the stage for what might have proven to be a real confrontation, if any of the Commissioner had decided to show their strength.[12]

"You'd think that this was a command performance and we were living in England," said Monica Gabrielle. "To me, the fact that it took so long to get the National Security Advisor in front of this Commission was just indicative [of] how deferential the media and the Commission itself had become to this White House. Let's remember what was happening here: Condi Rice was a public servant. She and her team, by all accounts, were asleep on the bridge when this country faced the greatest external threat to its security since the Cold War. And we were having to *beg* the White House to make her accountable. I was just praying that after all these months, the Commission wouldn't let her off the hook."

The Sunday before her testimony, the *New York Times* summarized the stakes in a Week in Review piece headlined "Bush's Credibility Now Rests On Her Shoulders."[13] A companion timeline noted that between May and June 2001 "at least 33 intercepted messages suggested an imminent terrorist attack"[14]

Richard Clarke had told the Commission that "the Bush Administration saw terrorism policy as important, but not urgent prior to 9/11." Rice countered in the media that in an August 2002 press briefing Clarke admitted that she had acted on his ideas. "So he can't have it both ways."[15] Did Clarke paint an accurate portrait of his boss as one who seemed out of touch with the bin Laden threat? Or was he simply a disgruntled employee, stung that Dr. Rice had demoted him? Those were the questions; and the Commission would now have its opportunity to get the answers.

The Moment of Truth

On the morning of April 8, Condoleezza Rice, in a beige suit with an American flag pin in her lapel, raised her hand and was sworn before a packed Commission hearing at the Hart Senate Office Building on Capitol Hill. There was no staff statement that morning; there didn't have to be. Dr. Rice *was* the hearing. A gifted child prodigy who played concert piano and entered college at the age of fifteen, Rice had lived much of her life in the public eye—moving on to become Provost of Stanford University, before joining the campaign of George W. Bush.[16] Now she was prepared for the performance of her career.

As they always did, the widows—who never fly—had driven down from Jersey and Manhattan. They had seated themselves behind her in the gallery to keep watch.

The testimony of the National Security Advisor turned out to be a watershed moment for the 9/11 Commission. The March hearings, energized by the testimony of Richard Clarke and those tough staff statements, suggested that the panel might have found its stride in its mission to pull back the layers and give the American public the full truth behind 9/11. With Rice's appearance, though, the Commission and its staff seemed to take on a newly defensive, almost protective tone, as if the revelations spilling forth were so damaging to both the Clinton and Bush presidencies that the Commissioners themselves felt they had to pull back.

The attitude of ex-Democratic Senator Bob Kerrey in his questioning of Rice set the tone. Only days before the hearing he had been quoted as if he was armed for bear: "My gosh," he said, "I think she was on every single network the day the commission opened its hearing this week attacking our witnesses."

But when Rice was sworn in and sat across from the Commissioners, Kerrey, behaved like a smitten schoolboy: "Let me say at the beginning, I'm very impressed, and indeed I'd go as far as to say *moved* by your story," he said, "the story of your life and what you've accomplished. It's quite extraordinary . . . and . . . I'm not sure had I been in your position . . . that I would have done things differently. I simply don't know."

Rice began by attempting to explain her statement of May 2002 that "I don't think anybody could have predicted that these people would take an airplane and slam it into the World Trade Center, take another one and slam it into the Pentagon; that they would try and use an airplane as a missile."[17]

"I probably should have said, 'I could not have imagined,'" explained Rice, "because within two days people started to come to me and say, 'Oh, but there were these reports in 1998 and 1999. The intelligence community did look at information about this.'"

Actually, there was substantial evidence from years earlier that terrorists were preparing to use planes as weapons of mass destruction. The Joint Inquiry had listed those twelve examples between 1994 and 2001, among them the Christmas 1994 hijacking of an Air France jumbo jet by Islamic radicals aligned with al Qaeda, which was aborted by French special forces police. But the plotters were overheard by passengers saying that their plan was to fly the big fuel laden Airbus into the Eiffel Tower.

There was also the Philippines evidence as set forth in *1000 Years for Revenge,* which Governor Kean had read. In January he had written to me that the Commission staff had read the book and "found it helpful." By March 15, when I testified before the Commission in New York, the staff was well aware of Col. Mendoza's discovery that the hijacking airplanes suicide plot had originated in 1994. But neither Chairman Kean nor any of the other Commissioners thought to challenge Rice on her offhand characterizations of timing.

Instead, she claimed that "this kind of analysis about the use of airplanes as weapons actually was never briefed to us." Blaming her own apparent lapse on others, she concluded that "perhaps the intelligence agencies thought that the sourcing was speculative."[18]

There was nothing speculative in the title of the Presidential Daily Briefing of August 6, 2001: "Bin Laden Determined to Strike in US."

Nonetheless, Rice insisted that "it did not warn of attacks inside the United States."

But former Watergate counsel Richard Ben-Veniste cross-examined Rice in what soon turned into a prickly exchange:

> BEN-VENISTE Isn't it a fact, Dr. Rice, that the August 6, PDB warned against possible attacks in this country? And I ask you whether you recall the title of that PDB?
>
> RICE I believe the title was, "Bin Laden Determined to Attack Inside the United States." Now the . . .
>
> BEN-VENISTE Thank you.
>
> RICE No, Mr. Ben-Veniste . . .
>
> BEN-VENISTE I will get into the . . .
>
> RICE I would like to finish my point here.
>
> BEN-VENISTE I didn't know there was a point.
>
> RICE Given that—you asked me whether or not it warned of attacks.
>
> BEN-VENISTE I asked you what the title was.
>
> RICE You said, did it not warn of attacks. It did not warn of attacks inside the United States. It was historical information based on old reporting. There was no new threat information. And it did not, in fact, warn of any coming attacks inside the United States.

In fact, the PDB did warn that FBI information indicated "patterns of suspicious activity in this country consistent with preparations for hijacking or other types of attacks, including recent surveillance of federal buildings in New York."

But the Commissioners weren't able to get into those specifics at the time Dr. Rice was in front of them, because the White House kept the Daily Briefing classified until two days *after* her appearance. With the glare of the national media on her, Rice brassed it out, insisting again that "this was not a warning. This was a historic memo."

She was almost finished when she made an effort to explain how the Bush Administration had missed the significance of the intelligence warnings in the months before 9/11: "The problem was that for a country that had not been attacked on its territory in a major way in almost two hundred years, there were a lot of structural impediments to those kinds of attacks."

Three days before that session, forensic researchers Angela Clemente and Dr. Stephen Dresch had sent their timeline to the Commission, outlining the intelligence from Greg Scarpa Jr., who had unearthed probative evidence of al Qaeda's involvement in the crash of TWA 800. It is unknown whether any of that material was circulated to the Commissioners at the time of Dr. Rice's appearance. But if it had been, not one of the Commissioners challenged her.

After more than two and a half hours of testimony, Rice was released. Incredibly, not one of the Commissioners asked her a single question about the performance of the White House on the day of 9/11.

Two days later, the full text of the August 6 PDB (with several small redactions) was made public. It revealed that six weeks before 9/11 the FBI was conducting "approximately 70 full field investigations throughout the U.S." that it considered "bin Laden-related. CIA and FBI are investigating a call to our Embassy in the UAE (United Arab Emirates) in May saying that a group of bin Laden supporters was in the U.S. planning attacks with explosives."

There was nothing historical about that. The PDB, which the White House had tried to hide from the public, contained *active threat intelligence.* Taken alone, it might not have been significant. But coupled with the Moussaoui report, the Phoenix memo, and the thirty-three "imminent attack" messages, signaling what the CIA director had called a "blinking red" warning, the Presidential Daily Briefing became part of a threat mosaic that no competent National Security Advisor could deny. Still, the Commissioners allowed Rice to leave the panel relatively unchallenged, and unscathed, as she faced the cameras with a straight face and uttered a sentiment that the President himself would echo five days later:

"I will tell you that if we had known that an attack was coming against the United States; that an attack was coming against New York and Washington, we would have moved heaven and earth to stop it."

16

CHECKING THE BUREAU SPIN

Staff Statement No. 9: Law Enforcement, Counterterrorism and Intelligence Collection in the United States Prior to 9/11

1*000 Years for Revenge* outlined dozens of missteps by the FBI, particularly its New York office, in the twelve years leading up to 9/11. This staff statement, which focuses on the Bureau, rightly notes that "New York was the 'Office of Origin' for the al Qaeda program and consequently where most of the FBI's institutional knowledge of al Qaeda resided." To get a full appreciation of this government's negligence in failing to stop the 9/11 attacks, it is necessary to dig deep inside the FBI's New York office. But this statement, like others to come, only scratches the surface.

As noted by the staff, "The FBI played the lead role in the government's domestic counterterrorism strategy before September 11." But in recounting the history of the NYO's involvement, the Commission hides the truth. As with the statement regarding Ramzi Yousef and KSM, the staff's characterization of the three most important plots involving Yousef, Shiekh Rahman, and the Manila cell are not easily reconciled with the facts on the public record.

The 9/11 Commission: ***"World Trade Center Bombing.*** On February 26, 1993, six people were killed and over a thousand injured when a truck bomb exploded in the basement of the World Trade Center. The FBI was able to identify the perpetrators of the attack as radical Islamists who were followers of the 'Blind Sheikh,' Omar Abdel Rahman. Through an international effort, the attack's mastermind, Abdul Basit Mahmoud Abdul Karim (better known by his alias, Ramzi Yousef), was brought back to the United States to stand trial, and he, like some of his co-conspirators, was convicted."

The facts: The staff fails to note that the NYO had been on to members of the Yousef-Abdel Rahman cell since the Calverton surveillance of 1989, or that FBI agent Nancy Floyd had inserted undercover asset Emad Salem into the cell, until he was forced to withdraw because of the actions of NYO ASAC Carson Dunbar.[1] They fail to add that Salem warned the Bureau to follow two of Yousef's key co-conspirators (Abouhalima and Salameh), who would have led agents directly to Yousef before he set the bomb.

The staff fails to note that the FBI had one of Yousef's core cell members, Abdel Rahman Yasin, in custody, but released him after he claimed innocence. Yasin then fled to Iraq, and eventually the United States issued a $25 million reward for his capture.[2] And though they reference "an international effort," the staff doesn't mention that Yousef was caught by DSS and DEA agents, or that after the FBI failed to canvas the Su Casa, Khalid Shaikh Mohammed escaped. Further, the staff, which later raises questions about Yousef's ties to al Qaeda, leaves out the fact that he was arrested in a bin Laden-controlled guesthouse.[3] All of these lapses are significant because in narrowing the scope of its investigation, the 9/11 Commission neglected to inform the public that considerable culpability for underestimating the growing al Qaeda threat occurred during the presidency of George H. W. Bush. It was on *his* watch that the FBI first missed a chance to capture Ramzi Yousef, the true "mastermind" of 9/11.

The 9/11 Commission: ***"Landmarks Plot.*** Later in 1993, the FBI disrupted the 'Day of Terror' plot which followers of Sheikh Rahman were in the midst of planning. Their plan was to blow up landmarks in the New York City area, including the Lincoln and Holland Tunnels, the George Washington Bridge, the United Nations, and the New York FBI Office.

The FBI was able to prevent this attack by reactivating a source who had previously infiltrated this particular cell."

The facts: The staff neglects to mention that the entire Day of Terror plot was exposed by Emad Salem, the source whom the Commission says was "reactivate[d]." In truth, Salem was so upset over his mistreatment by ASAC Carson Dunbar that he demanded $1.5 million from the Feds before going back undercover to do what he had been willing to do previously for five hundred dollars a week, his original pay.[4]

The staff doesn't mention the fact that it was Salem who provided the intelligence that led to the arrest in Egypt of key Yousef co-conspirator Mahmud Abouhalima ("the Red")[5] or that on multiple occasions in the course of the Landmarks sting Salem became frustrated with Carson Dunbar and threatened to pull out, only to be brought back into the fold by Special Agent Nancy Floyd, who had first recruited him.[6]

Finally, the staff leaves out any mention of the punitive measures taken against Agent Floyd by the Bureau's New York office. Even though she had recruited Salem (arguably the most important undercover asset to that point in the war on terror) rather than being rewarded, Floyd was subjected to a five-and-a-half-year internal affairs investigation, after which she was found guilty of being "insubordinate" to Carson Dunbar and suspended.[7]

Conversely, Dunbar, the ASAC who had almost blown the Day of Terror enterprise, was given the opportunity to take a position as superintendent of the New Jersey State Police, while transferring to an ATF no-show job that allowed him to maintain his federal pension.[8]

The 9/11 Commission: *"Manila Airlines Plot.* In January 1995 the Philippine police uncovered the plot to blow up twelve airplanes bound for the United States. Two of the perpetrators had also discussed the possibility of flying a small plane into the headquarters of the CIA. The FBI, working with the Philippines government, was able to determine that Ramzi Yousef was involved in this attack, as was Khalid Sheikh Mohammed, the eventual mastermind of the September 11, 2001 attacks."

The facts: As noted earlier, the "small plane into the CIA" scenario was discussed by Murad in the first stages of his interrogation by Col. Rodolfo Mendoza, at which point Yousef's co-conspirator, a commercial pilot trained in four U.S. flight schools, revealed the full details of the plot that was ultimately carried out on 9/11—a plot first planned in 1994 and not 1996, as the staff would go on to report in Staff Statement No. 16.

The Commission staff's almost off-handed report that "Yousef was involved in this attack" fails to indicate the depth of Yousef's genius or his position in the al Qaeda hierarchy. The Bojinka plot, funded by Osama bin Laden via his brother-in-law, was Ramzi Yousef's idea, and he personally designed the unique Casio-nitro "bomb trigger," which he successfully exploded on the PAL test flight in December 1994.

There was another significant lapse by the Commission staff as they sought to rewrite history regarding the African Embassy bombings:

The 9/11 Commission: ***"East Africa Embassy Bombings.*** On August 7, 1998, al Qaeda operatives bombed the U.S. Embassies in Kenya and Tanzania, in nearly simultaneous attacks. Twelve Americans and more than 200 Kenyans and Tanzanians were killed, and over 4,000 were injured. The FBI deployed hundreds of agents and other personnel to Africa to investigate the attacks. Usama Bin Laden and 22 other individuals were indicted for their role in these attacks. Four of these individuals were caught and convicted."

The facts: What the staff doesn't say is that the man who did the reconnaissance for the embassy bombings, and who took the pictures that Osama bin Laden personally used to place the trucks that carried the deadly bombs, was Ali Mohammed, the ex-Egyptian army officer-turned Green Beret who had trained the Calverton shooters under FBI surveillance in 1989.[9]

The report fails to note how Mohammed lived for years in San Francisco, serving as an FBI "informant" and passing on worthless intelligence to the Bureau as a double agent while he worked directly for bin Laden, training his bodyguards and spying on U.S. interests for al Qaeda.

The staff report does, however, reveal some troubling details that help to explain why the FBI seemed to be so far behind the eight ball in the wall on terror:

- Despite the fact that President Ronald Reagan had designated the Bureau as the lead domestic counterterrorism agency as far back as 1983, students at the FBI Academy received only three days of counterterrorism training in their four month course at Quantico.

• A 1999 internal review by the Investigative Services Division, which sought to measure the Bureau's intelligence analysis function, found that 66 percent of the FBI's analysts were not qualified to perform analytical duties.

On May 9, 2001, Attorney General John Ashcroft testified at a hearing on the U.S. counterterrorism effort. According to the Commission staff, Ashcroft claimed that "the Justice Department had no higher priority than to protect citizens from terrorist attacks." Yet on the very next day, May 10, the Justice Department issued guidelines for developing the fiscal 2002 budget. The priorities were reducing gun violence and the war on drugs. Dale Watson, the first head of the Bureau's counterterrorism division, told Commission staff members that when he saw those priorities "he fell out of his chair."

Finally, in examining the conduct of the FBI's New York office, there is no mention in this Staff Statement of the Bureau's role in setting up the third-party-calling "pass through" via Roma Corporation in 1996, which allowed Ramzi Yousef to contact cell members outside of his federal jail in New York—a system that, according to informant Greg Scarpa Jr., Yousef exploited. As the Joint Inquiry mentioned in its 2003 report, while in custody Yousef was able to contact KSM. Yet the 9/11 Commission makes no mention of this blunder—or of how the Scarpa intelligence put the FBI and Justice Department in a position to prevent the crash of TWA 800.

Staff Statement No. 10: Threats and Responses in 2001

This staff statement on the cacophony of threat reporting, which warned of an impending al Qaeda attack, begins with an important admission by the staff regarding National Security Director Condoleezza Rice: "Rice told us [Richard] Clarke and his Counterterrorism and Security Group (CSG) were 'the nerve center.'" The reader should keep this statement in mind during our consideration in Chapter 19 of Staff Statement No. 17, which recounts the defense failures on the day of 9/11. Despite his demotion by Rice, the statement underscores the importance of Clarke as the one effective "manager" of counterterrorism inside the White House.

The following passage from the Staff Statement expresses just how urgent the threat reporting was in the months leading up to 9/11:

> At Rice's request, on July 5 the CIA briefed Attorney General John Ashcroft on the al Qaeda threat, warning that a significant terrorist attack was imminent, and a strike could occur at any time. That same day, officials from domestic agencies, including the FAA, met with Clarke to discuss the current threat. Rice worked directly with Tenet on security issues for the G-8 summit. In addition to the individual reports, on July 11 top officials received a summary recapitulating the mass of al Qaeda-related threat reporting on several continents. Tenet told us that in his world 'the system was blinking red,' and by late July it could not have been any worse. Tenet told us he felt that President Bush and other officials grasped the urgency of what they were being told. On July 27 Clarke informed Rice and Hadley that the spike in signals intelligence about a near term attack had stopped. He urged keeping readiness high during the August vacation period.

But where was the FBI during all of this? The Commission staff found that, outside of the New York office, the Bureau personnel interviewed "did not recall a heightened sense of threat from al Qaeda within the United States." Those interviewed included an international terrorism squad supervisor in the Washington field office who said that not only did he and his squad take no special "steps or actions" to respond to the heightened threat in 2001, but they were entirely unaware of it.

Perhaps most disturbing was the staff's interview with acting FBI Director Tom Pickard, who met with Ashcroft once a week in the summer of 2001. Though it was his practice to brief the attorney general on terrorism threats, Ashcroft reportedly told Pickard after two meetings "that he did not want to hear this information anymore."

The Saudi Airlift

This staff statement is perhaps most notable for its reported exoneration of the Bush White House on charges that it had intervened in the evacuation of 142 people, mostly Saudi nationals, including members of the bin Laden family, in the days after the 9/11 attacks, when there were severe restrictions on air travel.

The charges first came to light in a September 2003 *Vanity Fair* story by Craig Unger, which excerpted part of his soon-to-be bestseller *House of Bush, House of Saud.*[10] In the book, Unger reported that in the first forty-eight hours after the 9/11 attacks, when "virtually no one could fly," the Saudi ambassador to the U.S. (and Bush family intimate) Prince Bandar bin Sultan bin Abdul Aziz, "was orchestrating the exodus of more than 140 Saudis scattered throughout the country, including members of Osama bin Laden's family." The incident, which caused an uproar and charges of special treatment, was featured prominently in Michael Moore's hit summer 2004 documentary *Fahrenheit 9/11.*

In his book, Unger reported that a specially configured 727 flew five bin Laden family members from Boston out of the country on September 18. The next day a flight arrived in Boston after picking up Saudis and bin Laden family members in Orlando, Florida, and Washington, D.C. At each step on the multicity exodus, wrote Unger, FBI officials were often in conflict with Bandar's chargé d'affaires at the Saudi Embassy in Washington, which was helping to expedite the exodus.

But the 9/11 Commission staff seemed determined to prove Unger wrong, by insisting that none of the flights took off during the immediate forty-eight hours after the 9/11 attacks when all U.S. air travel had been locked down.

"Of course the real issue here isn't the timing of the flights," says Patty Casazza, "but whether or not potential material witnesses to a mass murder should not have been detained in the U.S. for questioning. Who were they protecting?"

Despite compelling evidence of a decades-long relationship between the Bush family and the Saudis as set forth in Unger's book, the Commission staff went out of its way to separate the White House from the Saudi flight decision, placing the responsibility on counterterrorism crisis manager Richard Clarke. Consider this verbatim excerpt:

> THE SAUDI FLIGHTS
>
> National air space was closed on September 11. Fearing reprisals against Saudi nationals, the Saudi government asked for help in getting some of its citizens out of the country. We have not yet identified who they contacted for help. But we have found that the request came to the attention of Richard Clarke and that each of the flights we have studied was investigated by the FBI and dealt with in a professional manner prior to its departure. No commercial planes, including chartered flights, were permitted to fly into, out of, or within the United States until September 13, 2001. After the airspace reopened, six chartered flights with 142 people, mostly Saudi Arabian nationals, departed from the United States between September 14 and 24. One flight, the so-called Bin Laden flight, departed the United States on September 20 with 26 passengers, most of them relatives of Osama Bin Laden. We have found no credible evidence that any chartered flights of Saudi Arabian nationals departed the United States before the reopening of national airspace.
>
> The Saudi flights were screened by law enforcement officials, primarily the FBI, to ensure that people on these flights did not pose a threat to national security, and that nobody of interest to the FBI with regard to the 9/11 investigation was allowed to leave the country. Thirty of the 142 people on these flights were interviewed by the FBI, including 22 of the 26 people (23 passengers and 3 private security guards) on the Bin Laden flight. Many were asked detailed questions. None of the passengers stated that they had any recent contact with Usama Bin Laden or knew anything about terrorist activity.
>
> The FBI checked a variety of databases for information on the Bin Laden flight passengers and searched the aircraft. . . . The FBI has concluded that nobody was allowed to depart on these six flights who the FBI wanted to interview in connection with the 9/11 attacks, or who the FBI later concluded had any involvement in those attacks. To date, we have uncovered no evidence to contradict this conclusion.

But there were several factual errors in the staff statement, revealed only after Judicial Watch, a conservative watchdog group, obtained documents from the Department of Homeland Security. They showed that an additional 160 Saudis were allowed to leave on some fifty-five commercial flights, and that some had taken off as early as September 11, bringing the total number of Saudis in the airlift to more than three hundred.

Further, the Commission staff said nothing about the flight from Tampa, Florida, on September 13 of a Learjet 35 carrying a group of Saudis (including a member of the royal family) to Lexington, Kentucky, under the protection of an ex-FBI agent and retired Tampa police officer. That story, which Unger recounted in his book, was first reported in the *Tampa Tribune* in October 2001 by Kathy Steele, Brenda Kelly, and Elizabeth Lee Brown.[11] Then, following the defense of the Saudi airlift by the 9/11 Commission, reporter Jean Heller advanced the story further in a piece for the *St. Petersburg Times.*[12] Heller noted that officials of Tampa International Airport (TIA) had sent a confirmation of the Learjet flight to the 9/11 Commission staff following its publication of the staff statement:

"The Commission's general counsel, Daniel Marcus, contacted TIA in a letter dated May 25, more than five weeks after publication of the Staff Statement which missed the Tampa flight, requesting information about 'a chartered flight with six people, including a Saudi prince, that flew from Tampa, Florida on or about Sept. 13, 2001.' "

Unger had reported earlier that the Saudis had asked the Tampa PD to escort the flight, but the job was handed over to Dan Grossi, a former member of the department, working in private security. Grossi then recruited Manual Perez, a retired FBI agent, to accompany him. As Perez related the story to Unger in his book, "They got the approval somewhere. It must have come from the highest levels of government."

Not until the final report did the Commission see fit to mention the story of the Tampa flight, and even then it was in one of the thousands of footnotes printed in 7-point type at the back of the book.

In a *New York Times* piece by Unger on June 1, he disclosed that the commercial flights with the 160 additional Saudis "originated from more than 20 cities including Chicago, Dallas, Denver, Detroit and Houston."

This story is significant, and not just because it suggests that a group of wealthy Saudis and potential material witnesses got special treatment by the Bush administration. Whether Richard Clarke made the call or it was approved above him, no other interest group that we know of, U.S. or foreign, got anything close to that level of handling at a time when airports were either locked down or just beginning to return to normal.

Further, since fifteen of the nineteen hijackers were from Saudi Arabia, any potential fleeing Saudi national should have been subjected to a *heightened* level of scrutiny; especially members of the same family as the man who directed the attacks. But the staff statement admits that only thirty of the more than three hundred fleeing Saudis were ever questioned by the FBI—fewer than one in ten of the evacuees.

Unger's reporting, based in part on the enterprising work of local reporters, points up what we have attempted to demonstrate in this book: that the official version of the 9/11 story is being distorted by the very body set up to tell it truthfully and without consideration to special interests. An analysis of the work of thousands of reporters, as collated in the files of the Center for Cooperative Research, is critical to any serious assessment of the Commission's findings. While it may be impossible to judge the Commission's work when it comes to the vast amount of classified evidence that remains "in the black," we *can* evaluate the staff's findings against the public record. As we get closer to the events of September 11 itself, and the defense failures that left America exposed to attack, the need to separate truth from half-truth becomes even more acute.

17

"NOT A SINGLE PIECE OF PAPER"

Staff Statement No. 11: The Performance of the Intelligence Community

This Staff Statement opens with an admission that ought to make most American taxpayers cringe, especially after the failure of the U.S. intelligence community in the years leading up to 9/11: "The United States spends more on intelligence than most nations spend on national security as a whole. Most of this money is spent on intelligence collection, much of it on very expensive hardware, such as systems based in space. Most of the Intelligence Community's budget is spent in the Department of Defense, in part because the collection systems are mainly managed by agencies set up in that Department and in part because of the substantial intelligence organizations created to support the armed forces and the operations of the unified military commands around the world."

In the years before September 11, if the human side of intelligence gathering known as HUMINT failed us, then the hardware side was equally delinquent on 9/11 itself. The failures of NORAD and the FAA will be examined in Chapter 19, but first we'll focus on this staff statement, which contains a series of errors and omissions.

For example, under a heading marked "Analysis of a New Danger," which seeks to explain how the intelligence community underestimated the al Qaeda threat, the Commission staff claims that "While we now know that al Qaeda was formed in 1988, at the end of the Soviet occupation of Afghanistan, the Intelligence Community did not describe this organization, *at least in documents we have seen,** until 1999."[1]

But there were a significant number of documents describing al Qaeda in the possession of the FBI as early as 1996—documents in the form of FBI 302s or other briefing memos resulting from the interrogation of Jamal al Fadl. Al Fadl was the young Sudanese who worked for the murdered Mustafa Shalabi at the Alkifah Refugee Center at the al Farooq mosque in Brooklyn. In 1988, al Fadl traveled to Afghanistan and Pakistan where he was present at the virtual creation of al Qaeda.

Like a young Mafia member, al-Fadl swore a *bayat* or pledge to the new terror network. He later became so trusted by Osama bin Laden that the billionaire used him in Khartoum as a courier, front man, and link to Sudanese intelligence. But in 1996 al-Fadl got greedy. After embezzling some $110,000 in illicit commissions on the sale of goods from one of bin Laden's companies, he gave himself up to the United States and became known as Confidential Source 1 (CS-1).[2]

The CIA debriefed al Fadl for the first six months, then turned him over to the FBI, who nicknamed him "Junior." The FBI debriefing memos on al-Fadl provided a virtual road map of al Qaeda that the Feds later used to prosecute bin Laden in absentia in 2001. If al Fadl was handing over al Qaeda's inner secrets to the Bureau in 1996, then, how can the Commission staff credibly maintain that "the Intelligence Community did not describe" al Qaeda until 1999? Didn't Philip Zelikow or Jamie Gorelick have access to the classified al-Fadl papers? If not, why not? It's not as though the Commission never *heard* of al Fadl; in fact, two months later, on June 16, 2004, FBI Agent Mary Deborah Doran referred to him repeatedly in her testimony before the full panel.

* Italics added for emphasis.

We'll examine Doran's analysis of the work of the FBI's New York office in the next chapter. But at this point, in evaluating the truth of the Commission staff's contention in Statement No. 11 that the intelligence community didn't describe al Qaeda until 1999, the reader might consider Doran's admission on June 16 that by "December 1996, Junior was established as a FBI cooperating witness against al Qaeda. Information provided by Junior spurred a continuing effort to target and apprehend al Qaeda suspects."[3]

How could the Commission staff, which surely preinterviewed Special Agent Doran, have missed the fact that the Bureau was in possession of numerous al Fadl debriefing memos by late 1996?

By the time of its final report, the Commission had corrected the oversight in part: on page 79 of the report, it noted that al Fadl had become a "star informant for the United States."[4] But the final report never explains the staff's earlier conclusion that al Qaeda was missing from the U.S. intelligence community's lexicon until 1999.

Planes as Weapons: Missing the Joint Inquiry's Findings

Later in Staff Statement No. 11, the Commission staff appears again to have overlooked the import of the 800-plus-page report of the Joint Inquiry that they were supposed to build on. Earlier, we quoted verbatim the Joint Inquiry's description of twelve cases where planes were to be used as weapons in terrorist plots. To that list can now be added a thirteenth.

As yet another example of how an independent researcher with little or no budget and no subpoena power was able to come up with compelling evidence even the 9/11 Commission missed, consider the discovery made by John Berger, a Cambridge, Massachusetts-based investigator who operates a Web site called www.intelwire.com.

Berger was the first journalist to break the story of how Mohammed Jamal Khalifa, Osama bin Laden's brother-in-law, was actually held in U.S. custody through May 1995, when he was freed and deported to Jordan.[5] It was another case of the U.S. government expediting the exodus of a Saudi—in this instance, according to documents from the Philippines National Police, the financier of the Yousef-KSM Manila cell. When Yousef was arrested in Islamabad in February 1995, he was carrying Khalifa's business card.

John Berger and I first connected in the fall of 2003 after I discovered his remarkable website. We met in December in greater Boston, and the next month I sent him copies of the transcripts for both World Trade Center bombing trials, the Bojinka trial, and the Day of Terror trial, which had led to the conviction of the blind Sheikh.

Mining the trial transcripts, John discovered a 1993 plot by Islamic radicals tied to Sheikh Rahman to hijack an airplane and use it as a weapon. That plot, he learned, was directed against Egyptian president Hosni Mubarak and the U.S. embassy in Cairo. The plot was mentioned in the testimony of Emad Salem, Nancy Floyd's asset, who had gone back undercover for the FBI to break the Day of Terror plot. During the trial, Salem testified that Siddig Siddig Ali, one of the Sheikh's co-defendants, had asked him for help in a proposed plot in which a Sudanese air force pilot would bomb Mubarak's home from a plane and then eject before sending the aircraft crashing into the U.S. embassy.[6]

According to the transcript as quoted by Berger, Ali asked Salem to find "gaps in the air defense in Egypt," so that the pilot could "bomb the presidential house and then turn around, crash the plane into the American embassy after he eject[ed] himself out of the plane."

This information was part of the public record in a high-visibility Federal trial in the Southern District of New York, the FBI's bin Laden office of origin. Berger's discovery represented the thirteenth instance since 1993 in which Islamic terrorists contemplated using a hijacked aircraft as a weapon of mass destruction. But Staff Statement No. 11, in the spirit of Condoleezza Rice's testimony, seems to downplay the planes-as-weapons scenario. "Amid the thousands of threat reports," says the statement, "some mentioned aircraft in the years before 9/11.[7] The most prominent hijacking threat report came from a foreign government source in late 1998 and discussed a plan for hijacking a plane in order to gain hostages and bargain for the release of prisoners such as the 'Blind Sheikh.'" This instance was cited in the August 6, 2001, PDB published on page 308.

The staff statement also mentions briefly the Eiffel Tower hijacking plot from 1994 and the Murad "small plane into CIA" canard from 1995. But it fails to underscore how broadly known the concept of planes-as-weapons was inside the U.S. intelligence community.

Noting that the Counterterrorism Center (CTC) at the CIA "did not analyze how a hijacked aircraft or other explosives-laden aircraft might be used as a weapon," the staff statement singles out no one for blame. That may be because one of its authors was staff member Douglas MacEachin—who served as deputy director of intelligence at the CIA until 1995.

"I don't understand," says Lorie van Auken. "How does the Commission use a senior retired CIA official to evaluate the work of the CIA during his tenure? How can they possibly expect a transparent, objective analysis?"

Later in this book we'll examine exactly how many Commission staff members had similar conflicts of interest.

Staff Statement No. 12: Reforming Law Enforcement, Counterterrorism and Intelligence Collection in the United States

As protective as Commission staffer Douglas MacEachin was over the CIA in the last statement, FBI Director Robert Mueller was given kid-gloves treatment in this statement, co-authored in part by two Bureau veterans: Caroline Barnes, former senior counter-intelligence and counterterrorism analyst in the FBI's National Security Division, and Michael Jacobson, an ex-assistant general counsel and analyst in the same division.

While relatively tough on the Bureau during the Louis Freeh years, the Commission went easy on Mueller, a former Justice Department prosecutor, who took office just a week before the 9/11 attacks. And while he can't be faulted for the Bureau missteps *prior* to his tenure, the defensive posture Mueller adopted immediately *following* the attacks does merit closer attention.

Speaking at the Commonwealth Club in San Francisco seven months after 9/11, Mueller insisted that "the hijackers left no paper trail. In our investigation, we have not uncovered a single piece of paper—either here in the U.S. or in the treasure trove of information that has turned up in Afghanistan and elsewhere—that mentioned any aspect of the September 11th plot."[8] Testifying on Capitol hill two weeks later, Mueller repeated that claim, but took an even more defensive stance, suggesting that while the Minneapolis agents had done a "terrific job" in attempting to alert Headquarters to the threat from Zacarias Moussaoui, it was doubtful whether any further investigation would have led to the 9/11 plot. "Did we discern from that, there was a plot that could have led us to September 11th?" asked Mueller. "No. Could we have? I rather doubt it."[9]

Two weeks later, Coleen Rowley, a lawyer in the Minnesota office, sent a thirteen-page letter to Mueller expressing her "deep concern" that the director and other FBI leaders were "shading/skewing" the facts about the Bureau's failures leading up to 9/11.[10] Though she claimed "the term 'cover-up' would be too strong a characterization," Rowley complained that a supervisory special agent at Headquarters had been "consistently, almost deliberately thwarting the Minneapolis FBI agents' efforts" to investigate Zacarias Moussaoui in the weeks prior to the attacks.

Accusing the SSA of "deliberately sabotag[ing]" the Moussaoui case, Rowley wrote that some agents joked at the time that Headquarters must have "spies or moles" working for Osama bin Laden. That's how persistent they were in undermining the Bureau's counterterrorism efforts.

Warning that Mueller himself had not been "completely honest" about "the true reasons for the FBI's pre-September failures," Rowley predicted that "until we come clean and deal with causes, the Department of Justice will continue to experience problems fighting terrorism and fighting crime in general."

Rowley's candor made her an overnight heroine. She testified before the Senate Judiciary Committeem and was named one of three Persons of the Year by *Time* magazine for her courage as a whistleblower.

Meanwhile Mueller, who had initially kept Rowley's letter secret, got blasted by Senator Chuck Grassley (R-IA), the chief FBI oversight man on the Hill.

"A cover-up is not going to work," threatened Grassley.

Almost immediately, Mueller announced a sweeping series of FBI reforms. He pledged to reassign four hundred of the Bureau's 11,500 field agents from narcotics investigations to counterterrorism, and promised that four hundred new analysts would be added at the Bureau's fifty-six field offices and at Headquarters, including twenty-five from the CIA. The number of Joint Terrorism Task Forces, he said, would be expanded from thirty-five prior to 9/11 to sixty-six by 2003.[11]

These reforms weren't even announced by Mueller until more than half a year after 9/11, and their announcement was timed, in part, to serve as damage control after the rebuke he got for having minimized the dire warnings of an FBI whistleblower. But this is how the 9/11 Commission spun the story in Staff Statement No. 12:

"On September 4, 2001, Robert Mueller became the Director of the FBI. Soon after the attacks, Director Mueller began to announce and to implement an ambitious series of reforms aimed at, in his words, 'transforming the Bureau into an intelligence agency.'"

It's true that Director Mueller spoke publicly about "transforming the Bureau." But the first time he appears to have done so was in his May 8, 2002, Senate testimony,[12] and the next time was on June 18, 2003, when he testified before the House.[13] And while the staff statement took a few knocks at Mueller, noting that "many agents in the field were offended by the Director's statements that the FBI needs a new, proactive culture," the report was a relative valentine to the Bureau, declaring that "The FBI is widely regarded as one of the best post-event investigative agencies in the world."

Meanwhile, except for a brief reference in Staff Statement No. 9, the Commission staff virtually ignored the FBI's biggest ongoing embarrassment as it struggled to become a pro-active force effective at thwarting acts of terror in the post-9/11 era.

The Trilogy Project

Investigators for the Congressional Joint Inquiry had concluded that problems with the FBI's "paper driven" culture in the years leading up to

9/11 had prevented agents from sharing crucial intelligence that might have identified some of the nineteen hijackers and thwarted the attacks. Incredibly, in the twenty-four hours after the 9/11 attacks, even though it had identified all nineteen hijackers, the FBI was literally forced to send their mugshots to its fifty-six regional offices by overnight express mail, rather than e-mailing them as digital files—something a middle-school student with a scanner and a PC could have done without blinking.

And the system has not improved significantly since then. Staff Statement No. 9 reported that "The FBI received congressional approval in late 2000 for the 'Trilogy' project, a 36 month plan for improving its networks, systems and software. Former IBM executive Robert Dies, who was brought in by ex-Director Louis Freeh to consult, told the Commission that his goal was merely 'to get the car out of the garage.'"

The only other reference to Trilogy by the Commission was its admission that "the project was underway, but by no means fully implemented" at the time of 9/11.[14]

That turned out to be a gross understatement. In mid-May 2004, a review of the Trilogy project by the National Research Council (NRC) found that the system failed to support the Bureau's new counterterrorism mission and should be "redesigned." The project, launched in November 2000 with a $380 million price tag, has now grown into a $626 million white elephant. Director Mueller has repeatedly said that Trilogy is essential to the FBI's ability to "connect the dots," and prevent terror attacks.[15] But the NRC concluded that it was "not likely to be an adequate tool for counterterrorism analysis" and should be rebuilt "from scratch."[16]

A GAO report in March also noted that the FBI had gone through five information officers in the last two years since 9/11. The NRC study said that current Bureau plans to make a rapid changeover via Trilogy to a "virtual case file" system (VCF) was "highly risky" "and nearly guaranteed to cause mission critical failures and delays."

After that bleak NRC report, the Bureau asked for a re-review. By late June 2004, though, Bureau officials admitted that they wouldn't be able to deploy the much-needed computer system before the end of the year, and acknowledged that the VCF which would allow agents to easily share information, is two years behind schedule.[17]

"It's impossible to underscore how significant the failure of Trilogy has been in bringing the Bureau up to speed in the war on terror," says a retired Defense Intelligence Agency analyst who spoke on background. "Take a guy like Mahmud Abouhalima—one of the key players interacting with Yousef and the blind Sheikh back in '93. There are two or three different ways to spell his first name: Mahmoud, Makmud—then you've got his last name. It could be Abouhalima or Abu [meaning father of] Halima. The only way to be able to keep track of the thousands and thousands of al Qaeda members is by creating an enormous database that can cross reference and do link analysis. If the FBI doesn't get Trilogy online soon, they're going to continue to keep flying blind."

Classifying Ex Post Facto

Staff Statement No. 12 does acknowledge another big problem for the FBI in the war on terror: the lack of competent translators. "According to a recent report by the Department of Justice Inspector General, the FBI shortages of linguists have resulted in thousands of hours of audiotapes and pages of written material not being reviewed or translated in a timely manner. . . . The choice is between foregoing access to potentially relevant conversations and obtaining such conversations that remain untranslated."

But there's no mention in that staff statement of a smoldering scandal that has become a metaphor among 9/11 victims rights groups for the FBI's intransigence and slowness to change: the case of Sibel Edmonds. A former Bureau translator of Turkish descent, Edmonds was hired by the FBI shortly after the 9/11 attacks and terminated in the spring of 2002.[18] After a *60 Minutes* story in October of that year, in which she charged that the Bureau's translation capability was rife with incompetence and a lack of urgency, she was made the subject of an unprecedented gag order.

In June and July of 2002, staff members of the Senate Judiciary Committee had been briefed on Edmonds's allegations by the FBI. By March 2004, Sen. Patrick Leahy, the ranking Democrat on Judiciary, wrote to the DOJ saying that he still had "grave concerns" about the status of the FBI's ability to translate vital terrorism information.[19] But then in May the Justice

Department went to the extraordinary step of *retroactively classifying* the information Edmonds had given to Congress. In effect, they were declaring as *secret* a body of information that had been public for two years.

Edmonds, who had top secret clearance with the Bureau, now charges that the FBI had prior knowledge of the 9/11 attacks—proof, in fact, that al Qaeda was going to attack the United States with aircraft. In an interview with the British newspaper the *Independent,* Edmonds revealed that she testified to this in a closed-door session before the 9/11 Commission on February 11. "I gave [the commission] details of specific investigation files," she said, "the specific dates, specific target information, specific managers in charge of the investigation. I gave them everything so that they could go back and follow up. This is not hearsay. These things are documented."[20]

Kristen Breitweiser and the Jersey widows pushed hard to get the 9/11 Commission to address Edmonds's charges, but at the April hearing where Staff Statement 12 was presented and FBI chief Mueller testified, no specific questions were asked by the Commissioners.

The only reference to the issue came in a cryptic comment from Richard Ben-Veniste, himself a former federal prosecutor. "There's one area I want to put off to the side," he said, "and that's the area of FBI translators. I understand there are active investigations with respect to some of the allegations that have been made. I don't want to get into those facts now. I don't think it's appropriate." He then went on to declare that "The FBI is the finest law enforcement agency in the world, bar none."[21]

"Some of our group has met several times with Edmonds," said Lorie van Auken. "We think her claims are extremely credible. So much so, that some of our group hand-walked her in to testify before the 9/11 Commissioners." But in the end neither Edmonds nor her allegations about the FBI's prior warning of the 9/11 attacks were mentioned in any of the Commission's seventeen staff statements.

The only reference to Edmonds in the final report comes in a footnote to Chapter 3, which simply cites "A Review of the FBI's Actions in Connection with Allegations Raised by Contract Linguist Sibel Edmonds, July 1, 2004; Sibel Edmonds interview (Feb. 11, 2004)." Thus, any reader simply picking up the 9/11 Commission's official report would have no idea that an enormous scandal continues to fester inside the FBI and the Justice

Department over the issue of translators—or that Edmonds herself had told the Commission staff that the FBI had an advance warning of the 9/11 attacks.[22]

With the gag order in place and the Commission remaining silent, those charges of prior knowledge are difficult to confirm independently, but there was another example of an al Qaeda airplane hijack warning given to the FBI more than a year and a half before 9/11. Along with the Yousef-Scarpa intelligence, that case also got lost in what critics call the Bureau's "black hole."

It involved the strange but intriguing story of Niaz Khan.

The Walk-In Who Warned of the Plot

As mentioned briefly in the Joint Inquiry report, Khan, a British citizen of Pakistani descent, was a "walk-in" to the FBI's Newark office in March 2002. A former waiter in a Manchester, England, curry house,[23] the 29-year-old Khan had run up more than £15,000 in gambling debts when he was approached in the mill town of Oldham in 1999 by a pair of mysterious Islamic "brothers" who asked him if he knew who Osama bin Laden was.[24]

Khan told the Newark agents that men gave him £5,000 and flew him to Lahore, Pakistan, where he engaged in training by al Qaeda members on how to hijack airliners. Along with thirty other men, Khan said he learned to smuggle guns and weapons through security checkpoints, overpower crew members on board flights, and gain entry to cockpits. The slums of Lahore had been where Ramzi Yousef and Abdul Hakim Murad had first tested the Casio-nitro "bomb triggers" in 1994 that Yousef had designed for the Bojinka plot.

Khan later told Lisa Myers of NBC News that after a week of training he was given money to study flight operations and onboard security measures in a route that took him from Pakistan to Qatar, to London, then to Zurich and back to London. With a British passport, allowing him to enter the United States without a visa, he then flew to JFK, where he was supposed to link with an al-Qaeda contact. The man would reportedly be reportedly be waiting at a JFK taxi stand wearing a white skullcap. The code word for the cloak and dagger operation was supposed to be "Babu Khan." But the five-foot three-inch degenerate gambler lost his nerve and took off instead for Atlantic City, where he blew the rest of his al Qaeda money in the casinos.

Fearful of being hunted down by the terrorists, Khan says he turned himself in to the Newark office of the FBI. The Bureau confirmed that they questioned him for more than three weeks, and that he passed two polygraph tests. Khan told the agents that he was supposed to meet "five or six persons in the United States," some of whom would be pilots, "who had been instructed to take over a plane, fly to Afghanistan, or, if they could not make it there, blow the plane up."[25]

Incredibly, some of Yousef's kites to Greg Scarpa Jr., in 1996, verified by FBI 302s, confirmed that same kind of hijacking-flight-to-Afghanistan-plane-detonation scenario.

On September 9, 1996, four days after his conviction in the Bojinka case, the FBI confirmed that Yousef told Scarpa, "BOJINGA [his alias for bin Laden] contacted him during his trial and said that his people" were "going to prepare to hijack an airplane and kidnap a United States Ambassador at the time, if he is convicted. . . . YOUSEF stated that the hijacking could include more than one airplane [and that the airplanes would be] forced to land in either Afghanistan, Sudan or Yemen."*[26]

In another FBI 302 in late December 1996,[27] the Bureau noted that when Scarpa suggested to Yousef that he "give up 'BOJINGA' or BIN LADEN, YOUSEF responded that the Government would never go after BIN LADEN because the Government knows that within one week of capturing BIN LADEN twelve U.S. airplanes would be blown up."†

By the spring of 2002, nobody in the FBI's Newark office seemed to know about the 1996 Yousef-Scarpa intelligence—or, if they did, to have connected the dots.

*See FBI 302 on p. 306.

†See FBI 302 on p. 307.

Still, one ex-FBI official told NBC News that Newark agents believed Khan and tried to follow up on his lead, but then "word came from headquarters, saying, 'return him to London and forget about it.'"

Joseph Billy Jr., the agent in charge of the Newark office, said that Khan's claims were taken seriously. "An investigation was done on this matter when he came to us," Billy told the AP, "Nothing was discounted. We spent several weeks with him around the clock trying to verify the information that he gave us. None of the information that he gave us was ever able to be confirmed or denied."[28]

Unable to link Khan's charges to any ongoing crime, the Bureau deported him back to England in the spring of 2000 and turned him over to the British authorities. Myers reported that New Scotland Yard cut Khan loose after questioning him for only two hours.

Patty Casazza, whose husband John died in the 9/11 attacks, called the FBI's handling of the Khan case "another brand of negligence." "How many warnings do you have to have until news of a hijacking is to be deemed credible?" she asks.[29]

Again, all of these details surfaced in the mainstream press—and none of them appeared in any of the staff statements of the 9/11 Commission. Lisa Myers broke the story in the United States on June 3 after it had been in two British papers, and it made national headlines here. But not one of the Commissioners thought to raise the issue in the last series of hearings June 16 and 17 in Washington, even though the hearing on the 16th included the testimony of Special Agent Doran from the FBI's New York office.

On 9/11 Commission Comparing the FDNY and NYPD to "Boy Scouts"

The Commission's eleventh public hearing took place in New York City on May 18–19, and dealt with New York City's preparedness to deal with the attacks of 9/11. The first day of the two-day session was most notable for Commissioner John Lehman's attack on the alleged lack of communication on 9/11 between the New York City police and fire departments. "I think the command and control and communications of this city's public service is a scandal," said Lehman, on whose watch as Navy Secretary the 1983 terrorist bombing of the Marine barracks in Beirut took place. He went on to say that the city's disaster-response plans were "not worthy of the Boy Scouts, let alone this great city."[30]

Thomas Von Essen, the former FDNY commissioner, whose department lost 343 men that day, shot back angrily, "You make it sound like everything went wrong on September 11. I think it's outrageous that you make a statement like that."

The hearings also produced the assertion by former mayor Rudolph Giuliani that even if he had been fully briefed by the Bush administration about the dire terror warnings throughout the summer of 2001, he would not have acted any differently in terms of altering the city's preparedness in the days leading up to the attacks. "I can't tell you we would have done anything differently," he declared.[31]

Because this book focuses largely on issues of national security and defense, we'll avoid an analysis of those hearings and refer readers to the staff statements themselves, which can be found at http://www.9-11commission.gov/staff_statements.htm, and the transcripts of the testimony at http://www.nytimes.com/pages/world/worldspecial5.

The twelfth hearing, spread across two days in Washington on June 16–17, proved to be the most revealing of all—in terms what the Commission staff said about the performance of the White House, the Pentagon, and the FAA on the day of 9/11.

Some of the evidence the staff revealed was downright jaw-dropping. But in more than any other set of hearings to date, the Commission went out of its way to keep the public from the full truth. In the following chapters we'll endeavor to fill in the blanks.

18

THE "LOOSE NETWORK" BEHIND 9/11

The hearing that began in Washington on the morning of June 16, 2004, produced one of the Commission's most distorted pictures to date of the terrorists responsible for the 9/11 attacks.

Staff Statement No. 15: Overview of the Enemy

The statement, which purported to describe "The Roots of Al Qaeda," was delivered by staff member Douglas MacEachin, the man who served until 1995 as deputy director of intelligence (DDI) for the CIA. He was far from the only one on the Commission staff with a potential conflict of interest. We've already noted that staff director Philip Zelikow served on the Bush transition team and had co-authored a book with Condoleezza Rice, the Bush administration's national security advisor. Commissioner Jamie Gorelick was deputy attorney general under President Clinton at a time when the Justice Department ignored evidence of Yousef's role in the

9/11 plot and covered up his link to TWA 800. She had also served as general counsel to the Pentagon.

Gorelick and Zelikow were in familiar company. Nearly half of the Commission staff members had ties to the very agencies they were charged with examining. Of the seventy-five staffers listed on the Commission's Web site, nine worked for the Department of Justice as prosecutors or DOJ personnel, another six had worked for the CIA, and six others were FBI veterans. Four staffers had worked at the White House, three at the State Department, and five others at the Pentagon.

The Commission staff also included representatives of the INS and the NTSB (which clearly failed in the case of TWA 800), and one staff worker who served on a key intelligence oversight committee on Capitol Hill—a committee which, if it had done its job in years past, should have been alerted to the stunning series of intelligence failures that led up to the attacks.

An examination of the Commissioners' official bios, as published on their Web site, shows that two of them (Ben-Veniste and Thompson) were former Justice Department prosecutors who had worked closely with the FBI. Fred Fielding was a White House counsel to Richard Nixon and served on the Bush-Cheney transition team. John Lehman was a former secretary of the Navy under Ronald Reagan in 1983 when the U.S. Marine barracks in Beirut was destroyed by Islamic terrorists, killing 241 servicemen. Co-chairman Lee Hamilton, Slade Gorton, and Tim Roemer all served in Congress on committees with intelligence oversight functions. And former Senator Bob Kerrey, the Democrat who replaced Max Cleland, was once vice chairman of the Senate Intelligence Committee.[1] Known as a strong supporter of CIA director George Tenet,[2] Kerrey won the Medal of Honor after he lost part of a leg during his Navy Seal service in Vietnam. But he was also involved in a My Lai-like massacre of unarmed civilians, first exposed by the *New York Times* in 2001.[3] Critics on the far left felt that the scandal, described by one reporter as "a soul-searing fog of war mistake,"[4] subjected Kerrey to conflicts another Commissioner might not face.[5]

"In one way or another, they all have some baggage that makes them loyal to the defense, intelligence or justice establishment that failed to prevent these attacks," says Kyle Hence of the 9/11 Citizen's Watch group. "Or there were other conflicts like those of Gov. Kean, who was on the board of Amerada Hess, a major oil company with middle east interests. None of the Commissioners came into this process free of bias."

Widow Monica Gabrielle puts it bluntly: "How do they do an objective review of intelligence failures, when the investigative staff and the Commissioners themselves are so close to the agencies or interests they're supposed to audit?"

The "Loose Network" Behind 9/11

Now, on the morning of June 16, sitting erect with a straight face, former CIA DDI Douglas MacEachin reached a conclusion about Ramzi Yousef and his uncle Khalid Shaikh Mohammed that ran counter to the evidence in five separate federal trials:[6]

"Whether Bin Laden and his organization had roles in the 1993 attack on the World Trade Center and the thwarted Manila plot to blow up a dozen U.S. commercial aircraft in 1995 remains a matter of substantial uncertainty," he said.

"Ramzi Yousef, who was a lead operative in both plots, trained in camps in Afghanistan that were funded by Bin Laden and used to train many al Qaeda operatives. Whether he was then or later became a member of al Qaeda remains a matter of debate, but he was at a minimum part of a loose network of extremist Sunni Islamists who, like Bin Laden, began to focus their rage on the United States. Khalid Sheikh Mohammed, who provided some funding for Yousef in the 1993 WTC attack and was his operational partner in the Manila plot, later did join al Qaeda and masterminded the al Qaeda 9/11 attack. He was not, however, an al Qaeda member at the time of the Manila plot. A number of other individuals connected to the 1993 and 1995 plots or to some of the plotters either were or later became associates of Bin Laden. We have no conclusive evidence, however, that at the time of the plots any of them was operating under Bin Laden's direction."

In my previous book and this one we have offered an array of evidence documenting Ramzi Yousef's violent history as Osama bin Laden's point man on terror, from September 1991 through his capture in 1995, including declassified memos from the Philippines National Police and the testimony of PNP Col. Rodolfo Mendoza[7] citing Yousef's direct involvement in the planning for the 9/11 attacks, which were carried out by his uncle Khalid Shaikh Mohammed.

This book further provides evidence that even while Yousef was in federal jail he planned and triggered the downing of TWA Flight 800 in order to effect a mistrial in the Bojinka case. Documentary evidence from the PNP and a host of other sources[8] demonstrates that bin Laden's brother-in-law Mohammed Jamal Khalifa was the key financial link to the three Manila-based plots in 1994 and 1995 and that the Egyptian radical Sheikh Omar Abdel Rahman, beloved by the al Qaeda hierarchy, was both an operational leader and spiritual source who linked both attacks on the World Trade Center.[9]

The U.S. intelligence community was so aware of the ties among Ramzi Yousef, Sheikh Rahman, and Osama bin Laden that by August 1999 the Defense Intelligence Analysis Center (DIAC) at Bolling Air Force Base in Washington had produced a link chart showing a direct relationship between Yousef's 1993 WTC co-conspirators, the Sheikh's 1995 Day of Terror cell, Ali Mohammed, Wadih el-Hage, the African Embassy bombing cell, Osama bin Laden, and Dr. Ayman al-Zawahiri, the Egyptian who many intelligence analysts believe effectively runs the day-to-day operations of al Qaeda. The chart has been reproduced in Appendix III.

Further, the Staff's assertion that "Khalid Sheikh Mohammed . . . was not . . . an al Qaeda member at the time of the Manila plot" defies belief. Not only by the staff's own admission did KSM contribute to Yousef's 1993 World Trade Center bombing plot, but he set up bank accounts in the Philippines to launder al Qaeda money via Khalifa that funded the Bojinka, Pope, and 9/11 plots.[10] (See the PNP chart in Appendix III, pp. 310–11.)

Further, the Commission staff's description of al Qaeda as "a loose network" of extremists represents an outdated view that was contradicted moments later at the hearing in testimony by Patrick Fitzgerald, the current U.S. attorney in Chicago.

Fitzgerald, who ran the Organized Crime and Terrorism Unit in the SDNY that prosecuted all of the key al Qaeda cases, properly painted a picture of al Qaeda as a network designed around a hierarchical cell structure, with "many of the capabilities of a sophisticated intelligence organization."

It was the characterization of al Qaeda as "loosely organized" that caused FBI and CIA officials to underestimate the bin Laden threat for years.

For example, in May 2001, after Osama bin Laden was convicted in absentia in the U.S. District Court for the SDNY, *New York Times* reporter Benjamin Weiser filed a story with the headline "The Terror Verdict: The Organization; Trial Poked Holes in Image of bin Laden's Terror Group."[11] In it he quoted Larry C. Johnson, a former CIA officer who served as deputy director of the State Department's Office of Counterterrorism under George H. W. Bush and Bill Clinton.

"To listen to some of the news reports a year or two ago," said Johnson, "you would think bin Laden was running a top Fortune 500 multinational company—people everywhere, links everywhere. What the evidence at trial has correctly portrayed is that it's really a loose amalgam of people with a shared ideology, but a very limited direction." Weiser reported that "while Mr. Bin Laden may be a global menace, his group, Al Qaeda, was at times slipshod [and] torn by inner strife."

After the invasion of Afghanistan in mid-November 2001, when bin Laden's key lieutenant Mohammed Atef was killed, *Times* reporter James Risen quoted "a senior American law enforcement official" as saying "al Qaeda is on the run in Afghanistan, the noose is tightening around bin Laden, and I think what will happen is that Al Qaeda is going to fragment and that will make big-time operations more difficult."[12]

Four months later, an al Qaeda-suspected bomb exploded in a Tunisian synagogue, killing nineteen.[13] The following month, an al Qaeda car bomb in Karachi took fourteen lives.[14] The next month, in the same city, an al Qaeda-placed bomb detonated outside of the U.S. consulate, killing twelve.[15] In October, 2002 two hundred and two people died in Bali in a bombing directly tied to Riduan Ismuddin, aka "Hambali," the Indonesian cleric who had attended the January 2000 Kuala Lumpur 9/11 planning session and was on the board of Konsonjaya, the Malaysian front company that funded the Yousef-KSM Manila cell in 1995.[16]

In May 2003, al Qaeda suicide bombers killed thirty-four, including eight Americans, in a housing compound in Riyadh, Saudi Arabia. Days later, as a measurement of al Qaeda's worldwide reach, twenty-four people died in four separate bombs in Casablanca, Morocco. On August 8, a suicide car bomb killed twelve and injured 150 at the Marriott Hotel in Jarkarta. It was another attack linked to Hambali.[17] By November al Qaeda-suspected bombs had struck another housing compound in Riyadh, killing seventeen.[18]

The same month, twenty-five people were killed by bombs linked to al Qaeda in Turkey.[19] The next week the British bank in Istanbul was bombed.

The largest al Qaeda attack since Bali happened in March 2004, when explosive devices in backpacks ripped through the Madrid train station, killing 191.[20] Over two days in May, al Qaeda-suspected terrorists also attacked the offices of a Saudi oil company in Khobar, leaving twenty-two dead,[21] and as early as February 2004 evidence had surfaced suggesting that al Qaeda operatives—who had effectively stayed away from Iraq during Saddam Hussein's reign—had *now* infiltrated the country to promote conflict between Sunnis and Shiites in what has become a deadly and protracted uprising.[22]

The demise of Osama bin Laden and his network has been predicted over and over again by U.S. intelligence and defense officials, but the organization remains a tightly knit, disciplined force, capable of delivering mass casualties across the globe. Nonetheless, the 9/11 Commission's Staff Statement ends by offering the distinctly unpersuasive assessment that "al Qaeda today is more a loose collection of regional networks with a greatly weakened central organization."[23]

Staff Statement No. 16: Outline of the 9/11 Plot

Under the subtitle "Plot Overview," the statement begins by declaring that "The idea for the September 11th attacks appears to have originated with a veteran jihadist named Khalid Shaikh Mohammed (KSM)." After a brief description of KSM's contribution to his nephew's 1993 WTC bombing plot, and his role with Yousef in the Bojinka operation, the report remarks—without telling the real story—that "KSM managed to elude capture following his January 1996 indictment for his role in the plot."

The statement overlooks the fact that an FBI agent, Brad Garrett, missed Mohammed at the Su Casa guesthouse in 1995, and that the fleeing KSM somehow had the wherewithal to talk to a reporter from *Time*. More important, though, it leaves out a crucial bit of information regarding KSM's indictment: namely, that it was sealed. For more than a year and a half, no one outside of a few people in the Justice Department and the FBI were aware of KSM or his role with Yousef in the Bojinka plot. The first time his name surfaced publicly was on the inside "jump" page of a *New York Times* story on January 10, 1998, which covered Ramzi Yousef's sentencing in the Trade Center bombing.

Announcing that U.S. Attorney Mary Jo White had unsealed the 1996 indictment, the piece identified KSM as "Khaled Shaikh Mohammad, whom prosecutors described as a possible relative of Mr. Yousef."[24] Ramzi himself had been arrested after a very public worldwide manhunt, when Istaique Parker, an Islamic South African Yousef had recruited, sought to collect the State Department's $2 million reward.

KSM had the same price on his head, but for more than eighteen months the Feds had kept the hunt for him secret. Why? It was one of the questions that remained unanswered in the research for my last book, and in preparing my March 2004 testimony before the 9/11 Commission, I cited it as one that needed to be addressed by the panel.

One man who should have been able to fill in the blanks was Dietrich Snell, a former Assistant U.S. Attorney in the SDNY who (along with Mike Garcia) had prosecuted Yousef's Bojinka case. Although Khalid Shaikh Mohammed was a key co-conspirator in the Bojinka plot, his name was missing from the 6,000-page transcript of the trial. If anybody knew why, it was Snell.

Under objective circumstances, Snell would have made an important *witness* before the Commission. But in the heavily conflicted world of the Commission staff, he was hired to be one of its senior attorneys and team leaders. And on March 15, when I showed up at the Commission's New York office to share what I had learned about Yousef's role with KSM in the 9/11 plot, the man assigned by Philip Zelikow to take my evidence was Dietrich Snell.

In the two and one half years I'd been probing al Qaeda, what I had learned from reading more than forty thousand pages of court transcripts and analyzing hundreds of formerly classified files was that Ramzi Yousef and his uncle were key al Qaeda operatives. Their Manila cell was funded directly by bin Laden via his brother-in-law Khalifa, who had been on his way to join them in Manila when he was arrested by U.S. authorities in San Francisco in late 1994. PNP Col. Rodolfo Mendoza had been so certain that they had planned the 9/11 plot beginning in 1994 that when he watched the South Tower of the World Trade Center fall on 9/11, the first words out of his mouth were, "They have done it. They have DONE IT!"[25]

But Staff Statement No. 16, co-written by Dietrich Snell, concluded that KSM didn't begin planning the 9/11 attacks until 1996. There was no mention of Yousef's involvement in the plot. In fact, the statement suggested that KSM had been distant from al Qaeda in 1995, when he crafted the Bojinka plot with Yousef, and that it was only in 1996, a year later, that he sought to "renew" his acquaintance with Osama bin Laden.[26]

If this take on the story were true, it would let the Justice Department and the FBI off the hook for ignoring the evidence presented to them by Mendoza in 1995—evidence showing that Yousef and KSM had, by then, sent a series of Islamic radicals to train in U.S. flight schools, and that they were planning to hijack airliners and fly them into six U.S. buildings on both coasts, including the WTC, the Pentagon, and CIA headquarters.

But in Staff Statement No. 16, Snell and his coworkers (sourcing KSM himself) declared that he originally envisioned ten targets including skyscrapers in California and Washington State. No mention was made of the Sears Tower, which Abdul Hakim Murad told Mendoza had been on Yousef's list, even though a *New York Times* piece on KSM's interrogation by the Feds, two and a half months after his capture in March 2003, confirmed that America's tallest building remained a prospective al Qaeda target.[27]

In the Commission's telling of the story, written by Snell and other staffers, the FBI and the Justice Department couldn't be held accountable for a plot spawned more than a year after Ramzi Yousef had been captured.

The problem is, that story defied the evidence. As Mendoza put it, why would Abdul Hakim Murad, Yousef's cohort, have trained in four U.S. flight schools if the only plan was for him to plant the Casio-nitro bomb triggers on a series of planes and get off? This was a "separate plot" from Bojinka, Mendoza insisted.

In *1000 Years for Revenge,* we documented how during Yousef's Bojinka trial AUSAs Snell and Garcia called no fewer than eleven PNP officers and flew them to New York to testify. But Col. Mendoza, the chief PNP al Qaeda expert and interrogator of Yousef's accomplice Murad, wasn't one of them. In fact, twice during the trial Snell and Garcia allowed false or misleading testimony that distorted the true story of Murad's interrogation and the FBI's awareness of it.

During much of Murad's interrogation, Mendoza's chief assistant was Major Alberto Ferro.[28] Indeed Ferro testified during the Bojinka trial, but when asked who was interrogating Murad he said "I cannot really recall."[29]

When FBI agent Frank Pellegrino took the stand on August 6, 1996, he testified that he had arrived at Camp Crame, the site of Murad's interrogation, on January 17, 1995. At the time, Mendoza was grilling Murad only a few bungalows away. Yet the FBI agent swore under oath that he didn't even *know* the suspect was being questioned until months later, at the end of March 1995.[30] In terrorism cases, it's not unheard of for the Feds to let foreign officials get confessions from suspects before they're handed over and read their Miranda rights. But given that Murad was the partner of Ramzi Yousef, its difficult to believe that the FBI wasn't briefed at every stage of the interrogation by the PNP—especially since Pellegrino and FBI agent Thomas Donlon seemed so well-informed during their questioning of Murad on his flight back to the United States.[31]

Giving Unrecorded Testimony

Although it's anecdotal, my own experience with Dietrich Snell in relation to the 9/11 Commission is worthy of note for what it says about the inherent conflicts within the Commission staff. The major unanswered questions in *1000 Years for Revenge* were directly related to the conduct of

Justice Department lawyers like Snell in the months between May 1995, when Mendoza turned over his Murad file to the U.S. Embassy in Manila, and September 1996, when Snell and Garcia successfully convicted Yousef, Murad, and Shah in the Bojinka case:

- Why was the evidence that Ramzi Yousef conceived the 9/11 plot as far back as 1994 in the Philippines seemingly ignored by the FBI?
- Why did the Justice Department narrow the scope of the Bojinka trial of Yousef and two other al Qaeda cohorts in 1996—so that the name of the bomb maker's uncle, Khalid Shaikh Mohammed, "the mastermind of 9/11," wasn't mentioned in more than three months of testimony?
- Why did the FBI and Justice Department keep the hunt for KSM—a top al Qaeda operative with direct access to Osama bin Laden—secret for years, and fail to take advantage of the same public hue and cry that had brought his deadly nephew Ramzi to ground?

If I'd had access to Dietrich Snell while researching that first book, I would have asked him directly. Instead I got a chance to question him on the morning of March 15, 2004, when I was called to the 9/11 Commission's office at 26 Federal Plaza in Manhattan to "testify."

The word "testify" is set in quotes for good reason. When Snell led me into a conference room accompanied by staff member Marco Cordero (an FBI agent on loan to the Commission), there was no stenographer or recording device present. Then Snell sat down at a table across from me, pulled out a small pad, and proceeded to take notes. My source inside the Commission had warned mc about this: "People are watching the hearings and thinking that all of the witnesses we talk to are under oath and on the record," he said. "But it isn't true. More than ninety percent of the witness intake comes in 'informal' sessions, like the one they first had at the White House with Condi Rice." During these sessions investigators take notes and write up reports that are passed to the Commissioners for review. "But how does the Commission expect to make a complete record of events if it's so capricious?" asks lawyer and 9/11 widow Kristen Breitweiser.

It was Breitweiser and the other three Jersey girls who first suggested to the Commissioners that they review the findings in my book. During the October 14 hearing in Washington, the FSC members passed out copies of the book to the Commissioners, and Chairman Tom Kean read it over the Christmas holiday. On January 19 he wrote me to say that he had referred the FSC's request that I testify to Philip Zelikow. The staff director, in turn, put me in touch with Dietrich Snell.

Now, as I proceeded to give my "testimony" in the form of a written statement,* I stopped periodically to ask Snell some questions of my own.

"Why did you limit the Bojinka evidence at the trial?" I asked. "Did you know about Colonel Mendoza's revelations? Why did Agent [Frank] Pellegrino testify that he didn't know Yousef's partner Murad was being questioned at Camp Crame in Manila until late March of '95, when he'd been at the base for two months?"

Each time I asked a question, Snell would smile and say, "That's classified." Or "I can't discuss that." He was polite and cordial. At the end of my testimony he said that he would send me a "document request" asking for some of my research material.

I told him that I would forward what he needed, provided he didn't ask me to give up confidential sources. That had been one of my prerequisites to "testifying."

My other condition had been that my "testimony" become a part of the permanent record of the 9/11 Commission's proceedings. How that would now be accomplished without a stenographer present or a recording, I wasn't sure.

Two days after my testimony, Snell asked me to send him a series of supporting documents, including the transcript and a videotape of my interview with Mendoza on April 19, 2002, in the Philippines, along with various PNP documents.

On April 23, I assembled the package. But in the meantime I had received some of the evidence suggesting that the Justice Department, and

*Contained as Appendix I on pp. 273–293.

the SDNY office where Snell had worked, had been involved in a cover-up regarding the TWA investigation.

More than anything else, this new material on the Yousef-Scarpa intelligence pointed up the conflict of investigators like Dietrich Snell in acting on behalf of the Commission.

At that point I had more questions I wanted to ask him. In the course of the Bojinka trial, was he aware that his co-prosecutor Mike Garcia had received a death threat from Ramzi Yousef? If so, did he know about the FBI 302s from the Bureau's office across Federal Plaza that carried Yousef's warning? The terrorist had told Scarpa that if the trial began to turn against him, his cell would set a bomb aboard a plane to effect a mistrial. It was surely too tantalizing a coincidence to ignore.

I wondered if Snell had seen the July 24, 1996, FBI 302 a week after TWA Flight 800 went down. It read, "SCARPA advised that before the cell rotations MURAD stated that he feels that they may get a mistrial from the publicity surrounding the TWA explosion."[32]

More important, when FBI ADIC James Kallstrom did his about-face and decided to ignore the evidence of a crime, I wondered if Snell was aware of the ensuing DOJ cover-up?

Since he hadn't answered any questions at the time of my "testimony," I assumed I wouldn't get anywhere with this new line of inquiry. But for the sake of avoiding any possible conflict, I decided to bypass Snell and send the follow-up material directly to Governor Kean. In the cover letter I specifically cited my concerns about Snell's potential conflict of interest.

Five days later, the *New York Times* ran a complimentary story that focused on the former prosecutor, and quoted him as saying that before 9/11 he observed an incident in which one of Yousef's WTC bombing conspirators had threatened the Twin Towers:

"We're going to get them," said the terrorist. "We're going to get them."

This was, frankly, a new revelation. No Justice Department official had ever publicly admitted that the Yousef cell had predicted a return to finish the Towers, even though Ramzi's partner Murad had suggested as much to Federal agents during an interrogation and we had published the FBI 302 as an appendix to *1000 Years for Revenge*.[33] But this was evidence of prior knowledge by a DOJ lawyer of a threat to the WTC, years before the 9/11 attacks.

I wondered if it would show up in the Commission's final report—and it didn't. Nor was the reference in Staff Statement No. 16, which Snell read a section of at the hearing on June 16.

Weeks after 9/11, Dietrich Snell had granted an interview to Greg B. Smith, the *New York Daily News* reporter who had first broken the story of the Scarpa-Yousef intelligence initiative. Under the headline "On Clinton's Watch: Feds Nixed Deal For Plane Plot Tipoff," Smith reported on the early revelation from the Philippines National Police that Yousef's partner Murad had threatened to crash a small plane into the CIA.[34]

Asked about it, Snell told Smith, "It was not something that we focused on. It was something that he [Murad] said." We took seriously what he was telling us, but what we were focused on was the plot to blow up the twelve airliners. . . .

"The whole crux of Bojinka was to have timed explosions and the operatives to be off the flights and escaping," said Snell at the time. "That's a fundamental difference between what happened two weeks ago at the World Trade Center and Bojinka."

Snell was correct on that point. The 9/11 plot planned by Yousef, KSM, and Murad was fundamentally separate from Bojinka, as Mendoza had learned. But here in Smith's story was at least confirmation that Snell had been exposed to the PNP information.

The story ended this way: "Snell said he had no way to know whether Murad could have provided investigators with information that would be relevant to the probe of the Sept. 11 attack. 'I think it's pretty unlikely, but I don't know,' he said. 'I'd be guessing like everyone else.' "

Why does any of this matter? Because it's clear from the way the Commission was staffed—and the limited scope of its inquiry, which focused on the years from 1998 on—that entire areas of culpability by the U.S. Justice Department and the FBI have now been left out of the last official record of the causes behind the 9/11 attacks.

The only reference to my testimony to Dietrich Snell was this end note to chapter 5 in the final report:

After 9/11, some Philippine government officials claimed that while in Philippine custody in February 1995, KSM's Manila air plot co-conspirator Abdul Hakim Murad had confessed having discussed with Yousef the idea of attacking targets, including the World Trade Center, with hijacked commercial airliners flown by U.S.-trained Middle Eastern pilots. See Peter Lance, *1000 Years for Revenge: International Terrorism and the FBI—the Untold Story* (HarperCollins, 2003), pp. 278–280. In Murad's initial taped confession, he referred to an idea of crashing a plane into CIA headquarters. Lance gave us his copy of an apparent 1995 Philippine National Police document on an interrogation of Murad. That document reports Murad describing his idea of crashing a plane into CIA headquarters, but in this report Murad claims he was thinking of hijacking a commercial aircraft to do it, saying the idea had come up in a casual conversation with Yousef with no specific plan for its execution. We have seen no pre-9/11 evidence that Murad referred in interrogations to the training of other pilots, or referred in this casual conversation to targets other than the CIA. According to Lance, the Philippine police officer, who after 9/11 offered the much more elaborate account of Murad's statements reported in Lance's book, claims to have passed this added information to U.S. officials. But Lance states the Philippine officer declined to identify these officials. Peter Lance interview (Mar. 15, 2004). If such information was provided to a U.S. official, we have seen no indication that it was written down or disseminated within the U.S. government. Incidentally, KSM says he never discussed his idea for the planes operation with Murad, a person KSM regarded as a minor figure. Intelligence report, interrogation of KSM, Apr. 2, 2004.

This note, however, cannot be fully understood outside the context of Dietrich Snell's unique position as a former federal prosecutor, and the Commission's selective definition of what it considers "evidence." In its carefully worded declaration that "We have seen no pre-9/11 evidence that Murad referred in interrogations to targets other than the CIA," the Commission ignores the value of Mendoza's interview. Though we spoke to him in April 2002, after the attacks, Mendoza, the primary interrogator of Yousef's partner Murad, was an extremely credible witness. He wasn't just cited in *1000 Years for Revenge,* his allegations were widely reported by CNN in the fall of 2001.

But just as Snell and his partner Mike Garcia left Mendoza's name out of the Bojinka trial, the Commission staff fails to identify him here in the end note: he is referred to simply as "the Philippine officer." Because Mendoza refused to identify the U.S. officials he gave the evidence to, the staff seems to question the very validity of his information. But again, it should be noted that Dietrich Snell, who processed my evidence from

Mendoza, was one of the key Justice Department officials in a position to act on the colonel's 1995 warning about Yousef's "third plot," which became the 9/11 attack.

Further, the fact that the staff relies on KSM for proof that Murad (Yousef's trusted friend) was a "minor figure," is further evidence of the Commission's effort to minimize the importance of the true architect of the September 11 suicide-hijacking plot: Ramzi Yousef.

The Spin from the FBI's NYO

At that hearing on June 16, the Commission also heard from Special Agent Mary Deborah (Debbie) Doran. She described herself as "a street agent," working counterterrorism in the New York office of the FBI. The reader should keep in mind that New York was the same office that had failed to stop Ramzi Yousef in 1993, ignored evidence in 1995 from the Philippines that Yousef had begun to plan the 9/11 plot, and covered up evidence of Yousef's involvement in the crash of TWA 800.

Yet Special Agent Doran painted a far different picture.

"Let me begin by telling you that I am proud to be a Special Agent of the FBI, and that I am particularly proud of the work done by the Counterterrorism Division in New York," she said. "The FBI is extremely effective in putting together both criminal and intelligence cases all built upon information obtained through detailed and thorough investigations that are factually substantiated and corroborated."

None of the Commissioners—including Gov. Kean, who had read *1000 Years for Revenge*—thought to ask her about one of the NYO's most glaring omissions: the failure to follow up on evidence that in 1992 an Egyptian American accountant for the FDNY, who was an intimate of Sheikh Rahman, had obtained the blueprints of the WTC prior to the first bombing.

The accountant, Ahmed Amin Refai, had served as the Sheikh's personal bodyguard and translator at a 1993 immigration hearing in which the U.S. sought to deport the blind cleric. In 1998, Refai was caught by fire marshals in a series of lies after he tried to obtain a second ID to

Metrotech, the FDNY's highly secure headquarters, which contains the blueprints and plans of many of the city's most important buildings, including the World Trade Center.

Fire Marshal Ronnie Bucca, an Army reservist with top secret security clearance, had investigated the attempted breach with Fire Marshal Mel Hazel (now retired), and they determined that Refai was "a security risk." Bucca referred the matter to the FBI's New York office, but was told by a member of the NYPD-FBI Joint Terrorist Task Force that Refai had broken no federal laws, and they would not pursue the case.

Bucca, who had studied Islamic terrorism since 1993, had spent years warning anyone who would listen that Yousef's cell would return to New York "to finish the job." He died on September 11 on the 78th floor of the South Tower, while attempting to fight the fire caused by the crash of UA 175.

After his death, when the FDNY referred the Refai case back to the FBI, they balked a second time, but eventually an official of the Department of Homeland Security saw the FDNY file on Refai and sent it back to the Bureau once again.

In October 2002, the Refai case was reopened by the NYO, but a rookie agent was assigned to investigate it, and after several months it lapsed into an inactive status. Commissioner Kean was well aware of the incident, but during Special Agent Doran's testimony no one thought to ask her about the case.

The Refai case is a metaphor for the Commission's approach to investigating the Justice Department and the Bureau. In all of its public hearings, the staff and Commissioners never permitted a single witness outside of the government to offer testimony critical of the FBI.

My own research has revealed that the FBI's New York office was directly responsible for three enormous lapses—each of which, if rectified, might have helped agents stop the 9/11 plot years before it was finally executed.

First, they failed to capture Ramzi Yousef in the fall of 1992 as he prepped for the first World Trade Center attack. Since he was the true architect of 9/11, if the Bureau had captured him before the first bomb, they would have interdicted that plot as well.

Second, the FBI ignored probative evidence from the Philippines that the plot had its genesis in 1994—two years earlier than when the Commission now contends.

If they had acted on the evidence from the PNP back in 1995, the agents of the FBI's New York office might have taken a more aggressive role in hunting down Khalid Shaikh Mohammed, and if he had been captured, bin Laden would have lost the plot's key operational mastermind.

Finally, we now know that the FBI and DOJ were guilty of another crucial lapse of intelligence when they covered up the involvement of Yousef's cell in the downing of TWA 800—instead of pursuing an investigation that would have led them right to the heart of the 9/11 plot.

"What we've seen from the Staff Statements," says Kristen Breitweiser, "is that the Commission has ignored huge parts of the puzzle when it came to al Qaeda's early involvement in the United States—especially New York—focusing largely on the network's activities overseas."

But the Commission's selective accounting of the *intelligence* failures leading up to the 9/11 attacks *pales* in comparison to the unanswered questions raised by their report of the *defense* failures on the day of 9/11 itself.

19

EIGHTEEN MINUTES TO CALL NORAD

As a database for comparing the Commission's incomplete findings against the public record, we'll refer the reader to Paul Thompson's series of timelines contained at www.cooperativeresearch.org. A young American Stanford grad, Thompson, who now lives in New Zealand, began an exhaustive accounting of the public source record on 9/11 a year after the attacks. His vast history is backed up by articles that fill more than one hundred two-inch thick three-ring binders. Drawing on exclusively mainstream media sources, he brings together reports from ABC News, the *New York Times, Time* and *Newsweek*, the major broadcast and cable networks, and hundreds of other outlets, into one searchable database.[1]

As we'll see in the pages ahead, the official record of the events of the day of 9/11 itself changed in the thirty-one months between September 11 and June 17, 2004, when the Commission completed its last public hearing. Thompson's database allows us to reexamine the events as they were chronicled by print and electronic media sources during that period, providing an independent method of judging the Commission's performance.

Staff Statement No. 17: Improvising a Homeland Defense

The War Games

6:30 A.M. On the morning of 9/11, NEADS, NORAD's Northeast Air Defense Sector, is in the process of conducting a semi-annual war game involving airborne fighter jets. Called Vigilant Guardian, it is being monitored out of the NEADS center in Rome, New York, by Lt. Col. Dawne Deskins, the regional mission crew chief. NEADS protects the entire Northeast corridor of the United States, including the Washington, New York, and Boston metropolitan areas. At the time of the exercise, the entire NORAD chain of command is in a state of readiness, but the timing of the 9/11 attacks in the middle of the war game is curious. As *Aviation Week* and *Space Technology* later notes, "In retrospect, the exercise would prove to be a serendipitous enabler of a rapid military response to terrorist attacks on Sept. 11."[2]

In fact, Vigilant Guardian is one of *three* war games being conducted that morning. The second one involving fighter aircraft is a joint U.S.-Canadian exercise called Northern Vigilance.[3] And in perhaps the ultimate irony of September 11, the National Reconnaissance Office (NRO), which manages the nation's network of spy satellites, is conducting an exercise to test the NRO's response in the event that an aircraft should crash into its facility.[4]

The building is located four miles from Dulles Airport in Washington, the point of origin for one of the hijacked flights. The Vigilant Guardian exercise led to some confusion later at NEADS/NORAD, when personnel wondered if the real hijackings could be part of the game.

But in Staff Statement No. 17, the 9/11 Commission makes no mention of the three exercises. The sole reference to war games in the entire June 17 hearing comes from a question toward the end of the proceedings by former Congressman Tim Roemer.[5]

Unanswered Questions

- How many fighters were in the skies at the time of the first hijacking at 8:14 A.M.?

- What were their positions?
- Why weren't any of them tasked to respond to the attacks?
- Were any of the NEADS/NORAD officers controlling Vigilant Guardian confused by the actual reports of hijacking and did this in any way diminish their response time?
- Why did the Commission staff overlook these war games in its staff statement?

The Terrorists' Airport Access

6:30 A.M. At Boston's Logan Airport, an argument breaks out over a parking space involving five Middle Easterners and an unidentified man. When he reports the incident later, police discover a car rented by lead hijacker Mohammed Atta. It contains a pass allowing access to a restricted area of the airport from which two of the hijacked flights took off.[6] This was another mysterious incident overlooked by the Commission.

Unanswered Question

- How did the al Qaeda hijacker get that kind of restricted pass?

The Israelis' Early Warning

6:45 A.M. According to the *Washington Post* and *Ha'aretz*, the Hebrew language Israeli daily, at least two workers of Odigo, an Israeli-owned instant messaging company with offices two blocks from the Twin Towers, receive warnings about the impending attack two hours before the first plane hits the North Tower.[7] The Commission ignores this as well.

Unanswered Question

- Who gave these employees prior warnings of the attacks and how did they know what was about to transpire?

The FAA Delays Reporting on the Hijacking of AA Flight 11

7:59 A.M. Flight 11 takes off from Boston's Logan Airport, fourteen minutes after its scheduled departure.

8:14 A.M. Air traffic controllers order the pilot to turn right and he responds, but almost immediately he fails to obey a command to climb. At this point it appears that the plane has been taken over by Mohammed Atta and the other hijackers.

8:20 A.M. Boston flight control decides that Flight 11 has probably been hijacked. The protocol calls for NORAD to be notified *immediately,* but the Commission says that the FAA doesn't inform the North American Air Defense Command until 8:38 A.M., eighteen minutes later.[8]

Unanswered Question

- Why did it take the FAA so long to report the first hijacking to NORAD?

Conflicting Accounts of the Ability to Track the Flights

8:21 A.M. AA Flight 11 turns off its transponder, immediately "degrading the available information about the aircraft," according to the Commission. The staff admits that "Controllers at Centers rely on transponder signals and usually do not display primary radar returns on their scopes. But they can change the configuration of their radar scopes so they can see primary radar returns. In fact, the controllers did just that on 9/11 when the transponders were turned off in three of the four hijacked aircraft."

8:28 A.M. According to an ABC News story from September 6, 2002, Boston flight controller Mark Hodgkins recalled that "I watched the target of American 11 the whole way down."[9] This conflicts with other references in the Commission staff statement, suggesting that the four flights were "lost."[10]

8:40 A.M. Apparently, NEADS/NORAD's radar system is different from the FAA's, and it does not allow them to find AA 11. Boston ATC has to update NEADS on the flight's position periodically by phone until NEADS finally finds the flight—a few minutes before it crashes into the WTC.[11]

Unanswered Question

- How is it that the FAA was able to track AA 11 all the way into the WTC, but the North East Aid Defense Sector, with its sophisticated radar, couldn't locate the plane?

Eighteen Minutes to Notify NORAD

8:38 A.M. According to the Commission, Boston flight control contacts NEADS/NORAD at this time to inform them that AA 11 has been hijacked. Lt. Colonel Dawne Deskins, the mission crew chief for Vigilant Guardian, later says that initially she and "everybody" else at NEADS thought the call was part of the war game. She soon makes it clear to the NORAD command that this is not a drill.[12] Yet there is no mention of this in the Commission's final report, which offers only two obscure references to Lt. Col. Deskins in the end notes to chapter 1.[13]

Unanswered Question

- With fighters in the sky, why is Lt. Col. Deskins, the NEADS/NORAD coordinator of the war games, apparently so far out of the loop when a live hijacking is taking place?

The Missed Opportunity of the Atlantic City F-16s

Lt. Col. Deskins then informs Colonel Robert Marr, head of NEADS, who contacts Maj. General Larry Arnold at NORAD's command Center in Tyndall Air Force Base, Florida. He says, "Boss, I need to scramble [fighters at] Otis [Air National Guard Base]." Arnold authorizes the scramble.[14] But in a story filed December 5, 2003, Mike Kelly, a reporter for New Jersey's *Bergen Record,* notes that at that very moment two F-16s from the 177th Fighter Wing, based at Atlantic City International Airport in Pomona, are conducting bombing sortie exercises over the Pine Barrens in southern New Jersey. Though not armed with missiles, the planes are eight minutes from Lower Manhattan.

Lt. Luz Aponte, a spokeswoman for the 177th, tells Kelly that the wing commander was unaware that the two fighters were so badly needed. Although the existence of these fighters is barely referenced on the day of the Commission hearing, there is one reference in the staff statement that suggests they might have been able to interdict both crashes into the World Trade Center if notified in time by FAA.

The staff statement says that "In addition to making notifications within the FAA, Boston Center took the initiative, at 8:34, to contact the military through the FAA's Cape Cod facility. They also tried to obtain assistance from a former alert site in Atlantic City, unaware it had been phased out."

But while the former NORAD alert site was officially inactive due to cutbacks after the Cold War, the 177th was fully operational and had planes in the sky that day. It is unknown who in Atlantic City the FAA contacted, because the 9/11 Commission makes no mention of the 177th Fighter Wing or the two F-16s.

"But it couldn't be, because they weren't aware of them," says researcher Paul Thompson. "The *Bergen Record* story interviewed both Gov. Kean and John Farmer, one of the Commission's lead investigators."

"We want to know why jets at Pomona were decommissioned," Farmer told the *Record* reporter, who went on to ask why fighters at other bases closer to New York than Cape Cod weren't scrambled. "That's a big question," said Kean, also quoted in the story.

The point is, if the airborne F-16s had been notified by the FAA at 8:34, as the staff report says, they could have reached the Twin Towers by 8:45 A.M., a minute before AA 11 slammed into the North Tower. Even unarmed, and without a shoot-down order, they might have been able to take defensive action to prevent the big 767 from crashing into the tower. In any case, the fighters would certainly have been on patrol and able to interdict UA 175, which didn't hit the South Tower until 9:03 A.M.[15]

John Farmer, a former attorney general in New Jersey in the Kean administration, was considered by my source inside the Commission to be one of the most aggressive investigators on the staff. Yet, even though he read Staff Statement No. 17 at the June 17 hearing, Farmer made no mention of the very F-16s from the 177th he discussed in the *Bergen Record* story.

"I'm frankly stunned by this," says Lorie van Auken, whose husband, Ken, died when the North Tower was struck by AA 11. "If two fighters were only eight minutes away, the Commission should have done an exhaustive study on why they didn't get called. To leave them out of the official hearing record is unbelievable."

Unanswered Questions

- Why were the F-16s from the 177th Fighter Wing not contacted by FAA in enough time to thwart the crashes?
- Why did the Commission leave the entire incident out of its staff statement?

The Time It Took to Scramble the Otis Fighters

8:40 A.M. Otis Air National Guard (ANG) F-15 pilots Major Daniel Nash (codenamed Nasty) and Lt. Col. Timothy Duffy (codenamed Duff) are put on alert.[16] But they don't get into the air until thirteen minutes later, at: 8:53 A.M.[17] This is almost forty minutes after flight controllers lost contact with AA 11. Although their F-15s have a maximum airspeed of 1,875 mph,[18] they apparently fly toward New York City with airspeeds ranging from 500[19] to 900[20] mph.

If they had flown at even 1,200 mph, they could have reached New York in nine minutes and forty seconds and interdicted UA 175 just before it hit the South Tower at 9:03 A.M. But neither pilot is given a shootdown order to destroy the hijacked plane[21]—an order that could only have come from the president.[22]

Also, according to the Commission, the F-15s aren't directed by NORAD to New York City. Instead they are vectored toward military controlled airspace off of Long Island. There the F-15s remain in a holding pattern from 9:08 to 9:13 A.M. By the time they are ordered into Manhattan, UA 175 has already crashed into the seventy-eighth floor of the South Tower at 9:03 A.M.

Unanswered Questions

- Why were F-15s from Otis, 188 miles from New York City, scrambled when there were apparently other flighters closer to NYC?
- Why did it take pilots Nash and Duffy thirteen minutes to get airborne?
- Why didn't they fly to New York at maximum airspeed?
- Why were they first ordered to circle off shore?

A Second Plane Is Hijacked: Clearly an Attack Is in Progress

8:43 A.M. NORAD is notified that UA 175 has been hijacked.[23] NORAD learns instantly about the flight.[24] There's no doubt that some kind of attack on the United States is under way.

8:44 A.M. Secretary of Defense Donald Rumsfeld* is in his Pentagon office, ironically predicting that the country will be hit with a terrorist attack in the near future. "There will be another event," says Rumsfeld.[25] But with the country clearly under attack at this point, he remains in his office and doesn't go to the National Military Command Center (NMCC), the "war room" at the Pentagon, until 10:30 A.M.,[26] almost an hour and forty five minutes after this second hijacking and fifty-three minutes after the third hijacked flight, AA 77, hits the Pentagon itself at 9:37.[27]

* See picture, p. 269.

Unanswered Question

- Why does the Secretary of Defense apparently remain so far out of the loop during the worst attack against the U.S. homeland in history? Why doesn't he even enter the NMCC to take command until after his own building is struck?

The President Also Remains out of the Loop

8:46:40 A.M. AA Flight 11 slams into the WTC's North Tower.[28] At that moment, three F-16s assigned to Andrews Air Force Base, ten miles from Washington, are flying an air-to-ground training mission on a range in North Carolina, 207 miles away. Eventually they are recalled to Andrews Air Force base, ten miles from Washington, but they don't land there until some point after AA 77 crashes into the Pentagon at 9:37 A.M.[29]

President Bush will say in a speech that evening: "Immediately following the first attack, I implemented our government's emergency response plans"[30] But in fact the president at that moment is on his way to a photo opportunity at the Booker Elementary School in Sarasota, Florida. Press Secretary Ari Fleischer, traveling in the president's motorcade, learns of the first Trade Center crash at between 8:46–8:55, before Bush arrives at the school.[31] Once at the school, National Security Advisor Condoleezza Rice briefs the president by phone from the White House.[32] Rice later claims, "He said . . . it sounds like a terrible accident. Keep me informed."[33]

The Acting Joint Chiefs Chairman Also out of Touch

8:48 A.M. Meanwhile, Joint Chiefs of Staff chairman Gen. Hugh Shelton is out of contact, airborne across the Atlantic, and the vice chairman, Gen. Richard Myers, remains incommunicado, behind closed doors in a meeting on Capitol Hill with then-Sen. Max Cleland (D-GA). This despite the fact that, before entering the Senator's office, Myers sees a TV report of the plane crash into the WTC's North Tower. For the next fifty minutes, as additional planes are hijacked, the nation's chief military officer doesn't leave the meeting until the Pentagon is hit at 9:37. Apparently no one on the Senator's staff takes the initiative to interrupt the meeting. Myers later says that "nobody informed us" about the second WTC crash.

Unanswered Question

- In the midst of this defense crisis, how could the acting chairman of the Joint Chiefs remain behind closed doors on Capitol Hill, with no interruption and no input into the nation's military response?

8:56 A.M. AA Flight 77 has been hijacked.

9:03:11 A.M. UA 175 hits the WTC's South Tower.[34] As millions watch on TV, at least one hundred people are killed or injured on impact; another six hundred eventually die.

9:06 A.M. Chief of Staff Andrew Card enters the Florida classroom where the president is visiting students. He leans in and whispers, "A second plane hit the other tower, and America's under attack."*[35] The President looks startled, but remains listening to a story about a goat. He will stay in the room for another five to seven minutes. The 9/11 Commission accepts the president's explanation that he "felt he should project strength and calm until he could better understand what was happening."

Unanswered Questions

- Why would the commander in chief, the only one with the authority to order a shoot down of hijacked planes, even *enter* the classroom without being fully briefed on the first airplane crash into one of America's two largest office towers?

- Once he knew that the nation was being attacked, why didn't the president immediately vacate the school and head to Air Force One, the White House airborne command center?

*See picture, p. 269.

The Failure to Scramble Fighters to Protect Washington

9:03–9:06 A.M. The commander of the Pentagon's "war room" says that the National Military Command Center doesn't realize the hijackings are part of a coordinated attack until after UA 175 hits the South Tower. Yet NORAD knows of two hijackings by 8:56 A.M., when AA 77 is commandeered. A few minutes after 9:03, the Secret Service calls Andrews Air Force Base, ten miles from Washington. The base is notified to get F-16s armed and ready to fly. But missiles are still being loaded onto the F-16s when the Pentagon is hit more than half an hour later.[36]

Unanswered Questions

- Why did it take fighters at Andrews more than a half hour to get airborne, when F-16s from the D.C. Air National Guard (DCANG) were reportedly stationed at Andrews that day.

- Prior to 9/11, the DCANG's web site boasted of the unit's mission to protect the capital and "to provide combat units in the highest possible state of readiness." After 9/11 that mission statement was altered to state that the DCANG's job was merely to "provide peacetime command and control and administrative mission oversight to support customers, DCANG units, and NGB in achieving the highest levels of readiness."[37] Why was the mission statement changed?

The President Remains at the School

9:06–9:13 A.M. At this point four planes have been hijacked in an unprecedented attack on the nation. The president finally leaves the classroom, but remains in a nearby empty classroom preparing a speech to the nation,[38] which he will deliver at 9:29 A.M. The president doesn't leave the school until 9:34 A.M., more than thirty-eight minutes after his arrival.

Unanswered Question

- With at least three planes known to be hijacked and two already used as missiles, why does the president remain in the school to prepare a speech? At this point no one in the chain of command knows for sure if more planes haven't been commandeered, or what their targets might be.[39] If he wanted to talk to the nation, he could have done it aboard Air Force One. Mr. Bush's televised address makes Book Elementary School a highly visible target.

The Failure to Protect Washington

9:24 A.M. The FAA notifies NORAD that AA Flight 77 appears to be headed toward Washington, D. C.[40]

9:27 A.M. NORAD orders three F-16 fighters scrambled from Langley Air Force Base in Virginia. Langley is 129 miles from Washington. Ready aircraft at Andrews Air Force Base, ten miles away, are not scrambled.[41] The fact that both AA 11 and AA 77 are tracked by FAA, even when their transponders are shut off, proves that the government has the ability to follow each of the four hijacked aircraft on 9/11. In fact, Vice President Dick Cheney gets regular updates from the FAA as AA 77 bears down on Washington. He's first alerted when the hijacked plane is fifty miles away from the capital.

The *Wall Street Journal* has one explanation for the slow fighter response. In a story on March 22, 2004, they report that "Once they got in the air, the Langley fighters observed peacetime noise restrictions requiring that they fly more slowly than supersonic speed and take off over water, pointed away from Washington."[42]

Unanswered Questions

- If the FAA is monitoring AA 77, why do they wait to tell the White House until it's only fifty miles from D.C.?
- Why aren't D.C.-area fighters scrambled earlier to intercept it?

• What logic is there in scrambling planes 129 miles from the capital when it's now clear that the aircraft is heading toward D.C. and Andrews is only ten miles away?

• Traveling at speeds of 1,500 mph, the F-16s could have reached the Capital in a little over five minutes and interdicted AA 77 before it hit the Pentagon. If the slow response from the Langley fighters had to do with noise abatement rules, why didn't someone in the chain of command order the fighters to disregard them?

The Pentagon Crash

9:37 A.M. AA Flight 77 crashes into the Pentagon.[43] Accounts differ as to how far from Washington the F-16 fighters scrambled from Langley are when AA 77 crashes. NORAD originally claims that at the time of the crash the fighters are 105 miles away, despite having taken off seven minutes earlier.[44] The 9/11 Commission claims that at 9:36 A.M. NEADS discovers that AA 77 is only a few miles from the White House and is concerned to find that the fighters have headed east over the ocean. They are ordered to Washington immediately, but are still about 150 miles away. This is farther away than the base from which they took off.[45]

Unanswered Questions

• How is it possible that F-16s could have remained so far away from the capital throughout the time they were airborne? If they had flown at supersonic speed, they could have arrived in time to shoot down AA 77 before it hit the Pentagon.

• If the F-16s could get a radar fix from FAA or NORAD on the plane heading toward D.C., why couldn't they have taken it out with air-to-air missiles?

• Alternately, why weren't any ground-to-air missile batteries, ostensibly protecting Washington, put on alert to shoot down the incoming hijacked suicide airliner?

Air Force One Takes Off

9:56 A.M. The president leaves Sarasota, Florida. He later says that he doesn't make any major decisions about how to respond to the 9/11 attacks until he's in his airborne command center.[46] This means, in effect, that the commander in chief takes almost fifty minutes after being informed that "America is under attack" before taking action. It also conflicts with the president's statement that he "implemented" the government's emergency response plans "immediately following the first attack."

Unanswered Question

- Why? In their closed door session with the president and vice president, did the ten Commissioners of the 9/11 Commission ever press him on this, one of the greatest unanswered questions in the entire inquiry?

The Downing of UA Flight 93

After 9:56–10:06 A.M. Inside his White House bunker, a military aide informs Vice President Cheney, "There is a plane eighty miles out. There is a fighter in the area." He asks, "Should we engage?" Cheney reportedly answers "Yes."[47] By various accounts, an F-16 fighter (or fighters) take off in pursuit of UA 93,[48] even though, reportedly, they aren't armed. It is later alleged that they are supposed to crash into UA 93 if they can't get it to land.[49] In either case, as the fighter (or fighters) get nearer to UA 93, Cheney is asked twice more to confirm if the fighter(s) should engage. He responds "Yes" both times.[50]

9:58 A.M. Aboard UA 93, Todd Beamer ends a long phone call saying that the passengers plan "to jump" the hijacker in the back who has the bomb. He lets go of the phone, but leaves it connected. His famous last words are said to nearby passengers: "Are you ready, guys? Let's roll."[51] Sources claim the last thing heard on the cockpit voice recorder is the sound of wind—suggesting that the plane had been opened up, possibly by a missile.[52] But the Commission's final report suggests that the passengers never got into the cockpit, and that the terrorists themselves downed the plane.

10:03 A.M. According to the government, UA 93 crashes at 10:03,[53] just north of the Somerset County Airport, about eighty miles southeast of Pittsburgh, 124 miles or fifteen minutes from Washington, D.C. The recording appears to end a minute before the official crash time.

But a U.S. army seismic study concludes that, in fact, the crash happens at 10:06:05. The discrepancy is so puzzling that the *Philadelphia Daily News* does an investigation. In an article headlined "Three-Minute Discrepancy in Tape," the paper notes that "leading seismologists agree Flight 93 crashed at 10:06:05 A.M., give or take a couple of seconds, and government officials won't explain why they say the plane crashed at 10:03."[54]

Unanswered Questions

- Was UA 93 shot down?
- Why is there a discrepancy of three minutes and five seconds in the flight's official crash time and the army's seismic records?

A Delayed Change in the State of Readiness

10:10 A.M. All U.S. military forces are ordered to Defcon Three (or Defcon Delta), "The highest alert for the nuclear arsenal in 30 years."[55]

Unanswered Question

- Almost two hours have passed since the first plane was hijacked at 8:14 A.M. In the course of the morning as many as eleven planes are reported hijacked. The White House seems to have no idea of the dimensions of the attack. Why does it take so long for the Pentagon to order the nation into this higher state of readiness?

The 9/11 Panel Downplays the Role of Richard Clarke

As mentioned earlier, the 9/11 Commission was about to wind down in March 2004 when it was electrified by the testimony of terrorism czar Richard Clarke. As the *Washington Post* put it, "everyone seems to agree" that Clarke was the chief crisis manager on 9/11.[56] Even his boss,

National Security Advisor Condoleezza Rice, who challenged his testimony before the Commission, called him the "crisis management guy" for 9/11.[57]

Now, as we'll see, there are two scenarios for where our senior military leaders were on 9/11. One, the official version embraced by the 9/11 Commission, is that they were out of the loop—Rumsfeld in his office, Gen. Richard Myers, the acting Joint Chiefs chairman, locked in a meeting on Capitol Hill.

But Richard Clarke paints a different picture in his book, *Against All Enemies.* According to Clarke, a video conference is convened at 9:15 A.M., and most of the principals are involved in planning a response to the attacks, including Myers and Rumsfeld.

This account is taken from Paul Thompson's timeline of the events of 9/11 itself:

> **9:10 A.M. Clarke Directs Crisis Response Through Video Conference with Top Officials; 9/11 Commission and Others Barely Mention the Conference**
>
> Around this time, counterterrorism czar Richard Clarke reaches the Secure Video Conferencing Center next to the Situation Room in the West Wing of the White House. From there, he directs the response to the 9/11 attacks and stays in contact with other top officials through video links. On video are Defense Secretary Rumsfeld, CIA Director Tenet, FBI Director Mueller, FAA Administrator Jane Garvey, Deputy Attorney General Larry Thompson (filling in for the traveling Attorney General Ashcroft), Deputy Secretary of State Richard Armitage (filling in for the traveling Secretary of State Powell), and Vice-Chairman of the Joint Chiefs of Staff Richard Myers (filling in for the traveling Chairman Henry [Hugh] Shelton). National Security Advisor Rice is with Clarke, but she lets Clarke run the crisis response, deferring to his longer experience on terrorism matters. Clarke is also told by an aide, "We're on the line with NORAD, on an air threat conference call."[58]

After President Bush returns to Washington that night, he convenes a special meeting in the Situation Room. Clarke sits three chairs away from him, after chief of staff Andrew Card and CIA Director George Tenet.*

By Clarke's account, he was the primary crisis manager during the early hours of the attacks. Yet the 9/11 Commission never questions him

*See pictures, p. 269.

publicly about his performance on September 11, and devotes only two lines to him in its entire staff statement, dismissing his role:

> The White House Situation Room initiated a video teleconference, chaired by Richard Clarke. While important, it had no immediate effect on the emergency defense efforts.

Unanswered Questions

- Why would the 9/11 Commission seek to downplay Clarke's role as crisis manager on the day of the attacks? Further, whose version of events can we believe? Were Rumsfeld and Myers out of the loop, as the Commission suggestions, or where they in the thick of the response, as Clarke says?

"In the end, you have to ask yourself which is worse," says Monica Gabrielle, whose husband, Rich, lay injured and dying under a pile of debris on the seventy-eithth floor of the South Tower as the enormous building began to shudder and give way. "The idea that our nation's leaders were involved in every detail and *still* couldn't keep this from happening—or that they were AWOL? After all this time, we still don't have the full truth from the 9/11 Commission."[59]

20

"AMERICA'S UNDER ATTACK"

If there were any doubt as to why the 9/11 Commission Staff statements had been so incomplete, the mystery was cleared up in a UPI story by reporter Shaun Waterman on July 1, 2004. Revealing that the "Justice Department is reviewing the first chapters" of the Commission's final report prior to publication, Waterman reported that in February, White House Chief of Staff Andrew Card had established a process that cleared the staff statements through the White House.[1] The drafts of the first two chapters of the report, which was published on July 22, 2004, by W.W. Norton and Company, were now on their way to the DOJ to be vetted before release to the public.

"*Cleared* is too soft a word for what this is," says widow Monica Gabrielle. "From the beginning we were told that this would be an independent commission. We even named the web site of the Family Steering Committee 911independentcommission.org. But what this shows is that all along the White House, which stonewalled this Commission, would ultimately sanitize the final product."[2]

Kristen Breitweiser was more blunt. "This whole process smacks of censorship," she says. "We were never happy with the White House having the final edit. Of course national security must be protected, but the track record of this administration is that they've used the excuse of national security to hide information that is just embarrassing or inconvenient."

Staff Director Philip Zelikow, who brokered the deal with W.W. Norton to publish the report as a $10.00 trade paperback book, defended the White House approval process by noting that the Joint Inquiry was heavily criticized for redacting large portions of its 800+ page report as classified. "Nothing in this 'unclassified' report will hurt the national interest," said Zelikow.[3] Earlier, the staff director had come under fire for reportedly steering the book contract to his own publishing house.[4] In 2001, Zelikow, a professor at the University of Virginia's Miller Center, had co-authored a three-volume 1,536-page history of declassified recordings from President John F. Kennedy's White House meetings,[5] which Norton published. He also co-wrote another Norton history, *America and the East Asian Crisis: Memos to a President,*[6] with Robert B. Zoellick, who was appointed in 2001 to the cabinet-level post of U.S. Trade Representative in the Bush Administration.[7]

Even that association is controversial. According to his official bio, Ambassador Zoellick served with Secretary of State James A. Baker III during the administration of President George H. W. Bush. Baker, a Bush family retainer and senior partner in Baker Botts, the Houston law firm, is now defending a series of Saudi clients in the largest 9/11 lawsuit, brought by the South Carolina law firm of Motley-Rice.

"From its inception, as far as we can see, the 9/11 Commission was politically compromised," said Ron Motley in an interview with the author. "Not only are the Bush family's ties to the Saudis, extending back across three presidencies, virtually ignored, but here's the staff director, Zelikow, and his co-author *is a Jim Baker alumnus.* When the Commission went to such extraordinary lengths to let the Saudis off the hook, you have to wonder, what was going on behind the scenes?"[8]

Watered down or not, pre-orders for the Commission's final report were brisk, with the entire first printing of half a million copies reportedly selling out of its first run.[9] The *New York Times* called the Norton deal a "prestigious and potentially lucrative assignment."[10]

But while Commissioner Tim Roemer predicted in early May that the book would contain some "shocking information,"[11] co-Chairman

Lee Hamilton, Roemer's former House colleague, was talking about "toning down" the final version to avoid the possibility of a partisan split on the Commission.[12]

Apart from the prospect of White House censorship, "the worst part about this turned out to be that these Commissioners—virtually all Washington insiders—never really intended to find blame," says widow Patty Casazza, "and at this point we find that extremely troubling."

"These are very much people who are at heart of the establishment," says Ross Baker, a political scientist from Rutgers University, who monitored the Commission hearings. Calling the panel "probing but not aggressive," he says, "I think there's a very, very, strong disposition to avoid finger-pointing. It's very clear that they don't want to single out people for incompetence or not being vigilant enough."[13]

"They're missing the key elements that tell the story," snapped Lorie van Auken after the first day of the final hearing on June 16. Beverly Eckert went even further, flatly charging that "information is being concealed."[14] Bruce Dell, the father-in-law of one of the victims, called the hearing "a waste of time." Bringing a framed montage with thousands of tiny victims' pictures to the hearings, Dell told *New York Times* reporter Sheryl Gay Stolberg that he had given up hope of getting answers from the Commission. "I really just come to let them know that I'm watching," he said nodding toward the poster. "and let them look at these faces of the innocent people who were murdered."

Searching for Accountability

Even before that last hearing in early June, the Associated Press reported that the Commission's final report would "avoid finger pointing," in the hope of avoiding "partisan attacks." "We're going to say everything we need to say, but there's not going to be a political 'gotcha,'" said Commissioner Slade Gorton, a former Republican Senator from Washington state. "It's very important to be factual and leave major conclusions to the people of the United States. There are huge numbers of facts which are not in dispute."[15]

"There are also huge parts of the truth that the Commission has missed," said Mindy Kleinberg, commenting on Gorton's analysis. "I remember Governor Thompson, just before he questioned Rudy Giuliani at the New York hearings in May, saying something like 'this commission is not involved in a search for blame.' Well, how in God's name is this panel ever going to come with recommendations for change, when they have utterly, utterly failed to do that?"[16]

Even major publications seemed to have acquiesced. While some reporters—including William Bunch of the *Philadelphia Daily News,*[17] Eric Boehlet of Salon.com,[18] and Marie Cocco, a columnist for *Newsday*[19]—have been dogged in monitoring the Commission, and investigative reporters like Steven Komarow and Tom Squitieri of *USA Today* broke revelations that advanced the panel's findings,[20] certain news executives have treated the Commission with kid gloves.

On the June 17 edition of *Paula Zahn Now* on CNN, Michael Hirsch, a senior editor at *Newsweek,* called criticism of the Commission's reporting lapses concerning the day of 9/11 "Monday morning quarterbacking of a fairly useless kind."[21] And in May, seven weeks after their own reporters broke a page-one story on NORAD's Vigilant Guardian drills, a *USA Today* editorial entitled "End 9/11 Panel Theatrics," said, "Getting down to work instead of playing the blame game and engaging in public theatrics is the best way" for the Commission "to complete the job."[22]

On the second page of the Preface to its final report, the 9/11 Commission stated flatly that "Our aim has not been to assign individual blame."

A Final Request from the Widows

And so the question remains: Who is guarding the guardians themselves? And the answer, perhaps appropriately, comes down to "the widows": the four Jersey Girls (Kristen Breitweiser, Mindy Kleinberg, Patty Casazza, and Lorie van Auken), along with Monica Gabrielle, Sally Reganhard, Beverly Eckert, and the other members of the Family Steering Committee: Carol Ashley, Mary Fetchet, mother of Bradley James Fetchet, 24, who worked on the 89th floor of the South Tower; Bill Harvey, whose wife, Sara Manely Harvey, 31, worked with Carol's daughter on the 93rd floor of the North Tower; Robin Wiener, who lost her brother Jeffrey, 33, on the 96th floor; and Carie Lemack, whose mother, Judy Laroque, was a passenger aboard American Airlines Flight 11.

On June 14, two days before the last hearing, already aware that the Commission's record would be vastly incomplete, the Family Steering Committee sent this letter to the Commissioners expressing their disappointment:

> Dear Commissioners,
>
> Recent press reports indicate that no one will be held accountable by you in your Final Report because the Commission does not want to appear partisan. With five Republicans and five Democrats on your Commission, the issue of partisanship in the final report should not be an issue. Unless you write two reports—one by Republicans and one by Democrats, how could your Final Report be considered partisan? In telling the full story of September 11th, how can there be no accountability? Is no one in our government responsible for ensuring our nation's homeland security?
>
> In January 2003, each of you assured us of your commitment and resolve in meeting your mandate. We anticipated public hearings with tough, incisive questions. The sort of questions that would yield lessons learned, accountability, and solid reforms. Yet, over the course of eighteen months, we have been disappointed and frustrated with the lack of tenacity in your questioning and with the degree of your deference to witnesses. Too many times, you failed to ask the many crucial questions. Moreover, your lack of apparent resolve in gaining unfettered access to any and all documentation—including documentation from the Executive Branch and the City of New York, has undoubtedly hobbled the fact-finding process of your investigation.
>
> In this final public hearing and in your Final Report, we urge you to fulfill the pledge of commitment and resolve you made to us in January 2003. We ask you to pose tough questions. We implore you to reveal the investigative threads—no matter where they may lead. Americans must learn and understand why our nation was so vulnerable to terrorism on the morning of 9/11, and why 19 Middle-eastern men were so successful in carrying out their murderous plot, leaving thousands dead.
>
> Every day, more and more information emerges about the actions and omissions of our government on and around the morning of 9/11. We, therefore yet again, respectfully request that you present a definitive timeline of 9/11 to the American public. Such a timeline would prove to be invaluable as a framework for future analysis, the immediate identification of discrepancies of fact surrounding

9/11 as evidenced in your staff reports, public hearings and the media, and finally, the ultimate evaluation of federal and local level agency performances during a terrorist attack.

Your goal is to write the definitive report about 9/11 and issue recommendations based upon your findings. It is critical that your data be complete and accurate. If it is questionable or incomplete, your analyses and credibility will be undermined. Chillingly, your recommendations might therefore be ignored. The latest press reports regarding the Commission's reexamination of the Saudi flights may be construed as an indication of your thoroughness. Or, it could be an ominous indication that your investigation at this late date is critically incomplete. We realize and appreciate that the commission staff is working very hard in the face of a rapidly approaching deadline, but in this last phase, it is imperative that no stone remain unturned. The comprehensive and transparent fulfillment of your mandate must not be shirked.

Your Final Report and the successful implementation of your recommendations are the legacy you will leave your children and grandchildren. Your names will be attached to the Final Report to stand the test of time. Throughout the annals of history you will be held accountable for your success or failure to fulfill your mandate—a mandate that required you to fully investigate the worst terrorist attack in all of this nation's history. Success will mean a more secure America. Failure may mean that September 11th is only the beginning.

Respectfully,

The Family Steering Committee for the
9/11 Independent Commission

The Commission never produced such a timeline. That job was assumed by Paul Thompson and the Center for Cooperative Research, and it's clear now that the need for an independent accounting is more critical than ever.

The "Stonewall" Continues

On July 9, 2004, the Senate Intelligence Committee issued a blistering 511-page report that proved, definitively, that the information Congress had relied on in support of the invasion of Iraq, was "either overstated or . . . not supported by the underlying intelligence"[23] when it came to allegations that Saddam Hussein possessed weapons of mass destruction.

Delivering 117 separate conclusions, the report laid the blame squarely on what it characterized as a dysfunctional intelligence system led by CIA Director George Tenet. Sensing the writing on the wall, Tenet had resigned in early June.[24] He was followed by James Pavitt, a 31-year veteran who had been deputy director of operations, the Agency's official spymaster.[25] And while some critics saw this first bloodletting as an early sign that the 9/11 Commission had inspired some measure of accountability, the day before the scathing Senate report became public Tenet was awarded the National Intelligence Distinguished Service Medal.[26]

Further, despite the Intelligence Committee's blistering appraisal of the faulty intelligence on WMDs that paved the way to war, less than a week after the release of the report Douglas Jehl of the *New York Times* reported that the White House and the CIA were still refusing to give the Committee a one-page summary that had been prepared for President Bush before the war, which "contains few of the qualifiers and none of the dissents spelled out in longer intelligence reviews."[27]

Again, as with the August 6, 2001, PDB, the American people were left to wonder: "What did the President know and when did he know it?" And once again the White House was, to use that term made famous during the Watergate scandal, "stonewalling."

"In determining what the president was told about the contents of the National Intelligence Estimate dealing with Iraq's WMDs, qualifiers and all, there is nothing clearer than this single page," said Senator Richard J. Durbin, an Illinois Democrat, in an "additional view" that was released as an addendum to the Senate report. But as of the date this book went to press, that single document remains "in the black."

"Why should we care about this?" asks Monica Gabrielle. "Because Americans are dying right now in Iraq based on bogus intelligence—just as my husband, Rich, died on 9/11 based on bogus intelligence. And people will keep dying until we fix this and get a system that isn't just secret, but accountable to all the people."[28]

"It wasn't more than a month ago," said Lorie Van Auken in the spring of 2004, "that we were shown how this administration is willing to just plain lie when it comes to the war on terror. That's when the State Department admitted they had cooked the books on the terror stats."

She was referring to a report released by the State Department in April[29] that claimed the number of terrorist incidents worldwide had dropped in 2003 to 190. If that was true, it would have represented the lowest level in more than thirty years, and a decline of 45 percent since President Bush took office.[30] At the time, senior Bush administration officials like Deputy Secretary of State Richard Armitage cited the report as "clear evidence that we are prevailing in the fight," and Cofer Black, the ambassador at large for counterterrorism, called it "good news." Black is the Bush administration official who had stated in January 2004 that the al Qaeda terror network's days were numbered, two months before the devastating bin Laden-sponsored Madrid train bombing that left 191 dead.[31]

But after Democratic congressman Henry Waxman (D-CA) sought a review of the report by the Congressional Research Service, Secretary Colin Powell was forced to admit that terrorism activity was actually at a twenty-year high.[32]

"We are still trying to determine what went wrong with the data and why we didn't catch it," said an embarrassed Powell.[33]

The corrected report finally admitted that the total number of terrorist incidents actually *rose* in 2003. There were 390 deaths, versus 307 in the first report. And the number of people killed injured or kidnapped rose from 60 percent of the total incidents in 2002, to 89 percent last year. The State Department's Richard Boucher finally swallowed hard and admitted that based "on the facts as we had them at the time; the facts that we had were wrong."[34]

"But what would have happened if Henry Waxman hadn't challenged this?" said Lorie van Auken. "We would have been in the dark again. And every day that you pick up the paper there's another report telling us how we've been lied to."[35]

She cited a GAO report from mid-July,[36] which found that the Department of Homeland Security's color-coded threat warning system was too vague and confusing to help local law enforcement and first responders prepare for potential terrorist attacks. In a survey of twenty-eight agencies in fifty-six states, the nonpartisan investigative arm of Congress found that those responding "generally indicated that they did not receive specific threat information and guidance, which they believed hindered their ability to determine and implement protective measures."[37]

Representative Christopher Cox, the California Republican who heads the House Homeland Security Committee, said that officials in charge of monitoring the threat system ought to "make it work better or get rid of it."[38]

Meanwhile, Homeland Security head Tom Ridge had amped up fears just days earlier by announcing that al Qaeda was planning a major attack on the United States this year. "What we know about this more recent information is that it is being directed from the seniormost levels of the Al Qaeda operation," a top Homeland Security official said at a briefing.[39]

"So who do we believe?" asks Patty Casazza. "Tom Ridge is the man who told everybody to go out and buy duct tape just before the invasion of Iraq, because of concern for the weapons of mass destruction that didn't exist. We know that al Qaeda is a serious threat, and there's little doubt that they'll hit us again. But these broad scattershot warnings do nothing to make us safer."

Four days after the Ridge warning, John McLaughlin, who had stepped in as acting head of the CIA after George Tenet's departure, told a congressional panel that the terror threat leading up to the November elections was "as serious a threat environment as I have seen since 9/11."[40] And New York City Police Commissioner Ray Kelly upped the panic level, declaring that "New York is the number one target."[41]

"The question is, are these new terror warnings the next step in another reduction in civil liberties, like we saw with the Patriot Act?" wonders Monica Gabrielle.[42]

The day after she raised that question, the Department of Homeland Security confirmed that the Bush administration had discussed the possibility of delaying the November presidential election if there is a terrorist attack.[43] The story, first broken by Michael Isikoff in *Newsweek,*[44] was advanced by the Voice of America News.

Within hours, Congressman Cox, who had upbraided the Homeland Security Department days earlier after the GAO report on the color coded warnings, was calling the election delay request a prudent effort to plan for what he called "doomsday scenarios."[45]

Guarding the Guardians

After more than two years researching the terror threat and the way our Government has responded to it, I have come away with grave concerns. The Senate Intelligence Report, and the most candid parts of the 9/11 Commission's final report, paint a devastating portrait of an intelligence gathering system that is in a state of chaos and denial. As evidenced by the false information passed to Congress and the public in the days leading up to the Iraq war, it is clear that faulty intelligence can have a devastating impact on our national security.

Senator Jay Rockefeller, the ranking Democrat on the Intelligence Committee, flatly stated that if he had known then what he knows now, he would *not* have voted for the war resolution that sent so many U.S. troops into harms way.[46] At the same time, while admitting the prewar intelligence was "flawed," the president stood by his decision to go to war.

The important question now is whether those intelligence lapses were the result of negligence or intent. There's little doubt that, a year and a half after the first U.S. bombs hit Baghdad, the war on terror has become a tremendous profit center for at least one U.S. company: Halliburton. The multinational defense and logistics contractor, headed until 2000 by Vice President Cheney, was awarded a $15.6 billion no-bid contract after the Iraqi invasion,[47] for which Cheney had been one of the leading advocates.

As early as the summer of 2002, *New York Times* reporters Jeff Gerth and Don Van Natta Jr. reported that Halliburton had profited significantly from the war on terror: "From building cells for detainees at Guantanamo Bay in Cuba to feeding American troops in Uzbekistan, the Pentagon is increasingly relying on a unit of Halliburton called KBR," they wrote, noting that Halliburton's KBR subsidiary was the "exclusive logistics supplier for both the Navy and the Army, providing services like cooking construction, power generation and fuel transportation. The contract recently won from the Army is for 10 years and has no lid on costs, the only logistical arrangement by the Army without an estimated cost."[48]

Within months of the invasion in March 2003, KBR had become the object of an investigation into $250 million in disputed bills.[49] At the same time, investigators in France were probing whether KBR made $180 million in illegal payments in the 1990s to win a contract to build a natural gas complex in Nigeria. Those payments reportedly occurred at a time when the vice president was still Halliburton's chief executive, following the acquisition of the W.M. Kellogg Company, a unit of Dresser Industries that was absorbed into Halliburton after its takeover in 1998.

Last October, Congressman Waxman, who exposed the bogus State Department terrorism report, discovered, along with Michigan Congressman John Dingell, that Halliburton was charging U.S. taxpayers exorbitant prices to import gasoline into Iraq.[50] At that time Waxman and Dingell informed the Bush administration that KBR was billing the U.S. army between $1.62 and $1.70 a gallon, while Iraqis were charged between four and fifteen cents at the pump.

"Although Iraq is the second largest oil reserve in the world, the U.S. taxpayer is, in effect, subsidizing over 90 percent of the cost of gasoline sold in Iraq," the lawmakers said. At that point last fall Halliburton had already been paid $1.4 billion for the work it had done to extinguish oil fires—fires that would never have broken out if not for the invasion.

"Again," says Mindy Kleinberg, "you have to ask, 'where's the accountability?' "

As this book goes to press in early August 2004, the death toll to Coalition troops in Iraq has topped one thousand since the start of the war. At least eight hundred and eight of those killed are American men and women,[51] and 657 of them have fallen to hostile fire—more than five times the number killed during the two-month combat operation. At the same time, the insurgency grows every day. In early July, American military officials admitted to Associated Press reporter Jim Krane that the protracted revolt is being led by well-armed Sunni guerillas who can call on loyalists to boost their combat strength to as high as twenty thousand. That number exceeds, by four to one, the estimated five thousand insurgents previously believed to be armed and in the streets.[52] And kidnappings, car bombings, and threatened beheadings now occur on a daily basis.

At the same time, the U.S. force on the ground in Iraq, which declined in the fall of 2003, has jumped back up to 138,000 after troop rotations in January.[53] And the prisoner torture scandal at Abu Ghraib prison, which has diminished U.S. standing in the Arab world, grinds on with little immediate hope of accountability for the Pentagon officials who may have sanctioned the abuse. In an editorial in late June noting that the White House had obstructed investigations by the Army and the Senate Armed Services Committee, the *New York Times* used the same term cited throughout the 9/11 investigation: "stonewalled."[54] By mid-July, the *Times* was reporting that Republicans on Capitol Hill have virtually ground any further inquiries into the Abu Ghraib scandal to a halt until after the election.[55]

The Commission Proposals Fall Short

Based on two years of extensive research on the growing al Qaeda threat, it seems clear that the invasion of Baghdad was a mistake of catastrophic dimensions. It toppled a dictator who, while threatening his own population, played no role in the 9/11 attacks, and posed no threat to U.S. security. In the "shock and awe" campaign that followed, the U.S.-led Coalition ripped open the country and allowed members of al Qaeda to pour in, uniting Sunnis and Shiites with Baathists whose throats they might have cut prior to the invasion. The United States is now locked in a protracted guerrilla war that has become a titanic drain on American lives and treasury.

This is not a political position. As the reader knows, the vast majority of the intelligence lapses and cover-up activities chronicled in this book took place during the Clinton years. But the war on terror as directed by the Bush White House has degenerated into a kind of darkened room, where the public is subjected to daily fears based on unsubstantiated intelligence from an administration that uses "national security" as an excuse for avoiding accountability. Our vast intelligence community, with its multibillion-dollar annual expenditures, has failed repeatedly to appreciate the threat, or to protect us from it.

One Final Unbiased Investigation

One positive step—one crack of light in that "darkened room"—would be the creation of a truly independent panel, a new investigative body made up of journalists, scholars, victims' family members, and concerned citizens. The goal would be the creation of a populist blue ribbon commission, charged with two goals: filling in the gaps left by the Joint Inquiry and 9/11 Commission reports, and the proposal of an immediate and specific blueprint for overhauling the U.S. intelligence and early warning systems that have failed to protect America from the spiraling al Qaeda threat. As Justice Cardozo once wrote, "sunlight is the best disinfectant." Our spy agencies have failed because they have been allowed to fester too long "in the black," void of any true congressional oversight.

The 9/11 Commission's final report issued a series of proposals for restructuring the intelligence community, key among them the creation of a cabinet level terrorism "czar" and a National Counterterrorism Center which would, in the words of Richard Clarke, "ensure that our fifteen or so intelligence agencies play well together." But Clarke himself—perhaps the single person in the best position to know—wrote within days of the report's release that these changes would be "purely incremental." Calling the 9/11 Commission Report "toothless," he assures that, "Had these changes been made six years ago, they would not have significantly altered the way we dealt with Al Qaeda; they certainly would not have prevented 9/11."[56]

While the Jersey Girls initially praised the Commission staff for the "tireless work in producing the Final Report," they remain guarded and cautious as they page their way through the report. Lorie van Auken, the most pointed of the group, expressed her feelings about the need for full disclosure this way:

I would never wish what happened to us on anyone. After three years I'm incredibly weary, but we have to get the full story now and quickly. This isn't just our fight. It's America's fight. It's a battle for truth and accountability; one where people aren't afraid to say that if it's broken, let's fix it. We cannot let this threat from outside destroy what we love about this country. I want my husband back so much I feel it every waking moment. But I also want my country back. We have to get this done. We have no other choice.[57]

AFTERWORD

Few areas within the Justice Department's purview are shrouded in as much secrecy as the investigation of cases dealing with organized crime and terrorism. This book was built on dozens of source interviews, an examination of court files that now number more than fifty thousand pages, thousands of news stories, and hundreds of pages of staff statements and hearing transcripts from the 9/11 Commission, including its final report. Still, much of the subject matter remain classified. In measuring the depth of the DOJ's cover-up of the Yousef-Scarpa intelligence and its impact on the TWA 800 investigation, I relied on interviews with retired FBI personnel and the vast body of documents generated in the Colombo War cases. But in the end, after all of the 302 and 209 memos were examined, a single document—a letter—told me more about what was at stake here than a dozen interviews or hearing transcripts. It wasn't so much what the letter said on its face, as what it *didn't* say, and what the forces were in the background that generated it. At a critical stage in the completion of this book, the receipt of this letter spurred me forward to pursue the truth about the TWA 800 crash.

Though I was already in possession of several sworn affidavits quoting Greg Scarpa Jr. on the alleged corrupt relationship between his father and Lin DeVecchio, and had also obtained virtually all of his Yousef kites and the FBI 302s that authenticated them, I wanted to meet Scarpa face to face. There is no substitute for a direct interview with a source, and clearly Scarpa Jr. was one of the most important confidential informants in the history of the FBI's pursuit of the war on terror.

Both Angela Clemente, who had met with him at the ADX prison in Florence, Colorado, and Sandra Harmon, a Los Angeles-based writer who had interviewed Greg Sr.'s girlfriend Linda Diana Schiro, encouraged the younger Scarpa to talk to me. So on July 1, 2004, I wrote him to request an interview, letting him know that I would follow the proper protocol and contact his counselor at the prison, one Mr. Arroyo. After I spoke with Arroyo by phone, he instructed me to send a request in writing to Warden Robert A. Hood, and I did so by fax that same day.

Typically, unless an inmate is a particular security risk or there is an ongoing legal matter, requests by reporters for interviews are routinely granted by the U.S. Bureau of Prisons. Yet five days later, on July 6, a letter came over my fax machine in reply to my request. It ranks as one of the most telling documents I have ever received in more than thirty-five years as a journalist.

On letterhead from the Department of Justice, of which the Bureau of Prisons is a part, Warden Hood wrote:

"The Bureau of Prisons makes every effort to accommodate media personnel as long as the request does not negatively affect the security and operation of our institutions. It is my determination, based on my sound correctional judgment, that to grant your request at this time could pose a risk to the internal security of this institution and to the safety of staff, inmates and members of the public. Therefore your request has been denied and you will not be permitted to conduct an interview with inmate Scarpa at this time."

The Supermax, which houses Unabomber Ted Kaczynski, Oklahoma City bomber Terry Nichols, and a host of other convicted terrorists, ranks as one of the world's most impenetrable prisons. One wonders what kind of security threat a single investigative journalist might pose, armed only with a reporter's notebook and a pen.

My first thought on reading this letter was, *What do they have to hide? What makes the Justice Department so worried about what Greg Scarpa Jr. might tell a newsman?*

If, as Federal prosecutor Valerie Caproni's staff had argued, Scarpa's Yousef intelligence was little more than "a hoax" or "a scam," then why should they be worried *what* he says and *to whom?* I wanted to ask

07/07/2004 09:34 7197845290 ADX WARDEN PAGE 02/02

U.S. Department of Justice

Federal Bureau of Prisons

United States Penitentiary
Administrative Maximum

Office of the Warden

P.O. Box 8500
Florence, CO 81226-8500

July 6, 2004

Peter Anthony Lance
c/o Regan Books
Harper Collins
18th Floor
10 East 53rd St.
New York, NY 10022

Dear Mr. Lance:

This is in response to your correspondence dated July 1, 2004, received by fax on July 2, 2004, in which you request permission to conduct an interview with inmate Gregory Scarpa.

The Bureau of Prisons makes every effort to accommodate media personnel as long as the request does not negatively affect the security and operation of our institutions. It is my determination, based on my sound correctional judgement, that to grant your request at this time could pose a risk to the internal security of this institution and to the safety of staff, inmates and members of the public. Therefore, your request has been denied and you will not be permitted to conduct an interview with inmate Scarpa at this time.

Please feel free to contact my Executive Assistant and Public Information Officer, Wendy Montgomery, at 719/784-9464, if you have any further questions regarding this letter.

Sincerely,

Robert A. Hood

Robert A. Hood
Warden

by fax (212)734-0056

Valerie Caproni that. As it turns out, she remains in a position of great authority within the Justice Department to this day. In 1998, after leaving the U.S. attorney's office in Brooklyn, where she had become Chief of the Criminal Division, Caproni went on to become the regional director of the Pacific Regional office of the SEC, and in August 2003, Director Robert Mueller named her general counsel to the FBI.[1] In effect, she's the Bureau's top lawyer.

I also wanted to talk to Patrick Fitzgerald,* the former head of the Organized Crime and Terrorism Unit in the SDNY. Not only did Fitzgerald oversee the Bojinka case, but he reportedly wrote a sealed affidavit in 1998 authenticating Scarpa's Yousef intelligence.[2] Like Caproni, Fitzgerald too had risen to a position of great power in the Justice Department. He was now U.S. Attorney for the Northern District of Illinois's Eastern Division in Chicago, and he was the last senior DOJ official to testify before the 9/11 Commission.[3] As we noted, however, despite the fact that Chairman Tom Kean and Co-Chairman Lee Hamilton had received Angela Clemente's Scarpa-Yousef material in early April, no one on the panel thought to question Mr. Fitzgerald about it when he testified on June 16.

As much as any DOJ officials, Caproni and Fitzgerald should both have recognized the quality of Greg Scarpa's intel from Yousef. I had hoped that they would tell me why Justice had decided that it was worthless.

But they too declined to talk to me for this book.[4]

My requests for a chance to question Chairman Tom Kean, staff director Philip Zelikow, and Commissioner Jamie Gorelick were also effectively turned down. In an e-mail to me on May 24, Commission media director Al Felzenberg wrote that Kean and Zelikow would not be available until after publication of the Final Report, and a follow-up letter in July reiterating my request and asking for an interview with Gorelick went unanswered.

Still, as an investigator, the letter from Warden Hood was the most chilling of all the responses I've had from anyone in this three-year investigation. It might not have been definitive proof of a Scarpa cover-up, but

*U.S. Attorney Fitzgerald is also serving as Special Prosecutor for the Justice Department in the investigation of media leaks regarding Valerie Plame, a CIA officer who is the wife of former ambassador Joseph C. Wilson.

the Warden's obvious overreaction told me that someone behind the scenes at Justice was worried.

It also occurred to me that in suggesting that my visit might post "a risk to the internal security" of the prison, Justice Department lawyers might be laying the groundwork for allegations of Patriot Act violations in my very research.

As soon as I got the letter, I copied all of my research and moved it to a safe place. I then sent copies of the manuscript to lawyers in two separate cities across the country. Was this paranoia? I don't think so.

DOJ Suppressed a Reporter on the TWA 800 Story Once Before

In 1997, the FBI seized the phone records of investigative reporter Jim Sanders, a former police officer who espoused the missile theory in the downing of TWA 800.[5] After Sanders and his wife Elizabeth, a TWA flight attendant, reportedly received samples from the plane wreckage and had it tested for the presence of missile fuel, they were arrested and charged by the Justice Department with receiving evidence from a crime scene. They were both indicted,[6] arraigned, and released on $50,000 bail.[7] After a trial in U.S. District Court on Long Island, they were found guilty. This despite the fact that, at the time of the trial, the FBI had already acquiesced to the NTSB's mechanical theory and the Calverton hanger from which the "evidence" came could no longer be construed as a "crime scene." Though they could have faced up to ten years in prison,[8] a fair-minded federal judge, citing Sanders's background as a journalist, limited their sentence to probation.[9]

Still, it was the kind of example that would chill most reporters.

In fact, TWA 800 is a story that has tarnished the careers of a number of respected journalists including former ABC News Paris bureau chief Pierre Salinger,[10] who also espoused the missile theory, and Kristina Borjesson, a former freelance producer for CBS News, who received a portion of the evidence Sanders had obtained from the Calverton hangar.

In her book *Into the Buzzsaw,* Borjesson writes that after the sample had arrived at CBS News in New York, the FBI contacted her producers.

She was then summoned to a meeting with a CBS lawyer, who told her that he'd received a call from Valerie Caproni. The same prosecutor who would later discredit Greg Scarpa Jr.'s evidence was threatening to haul her in front of a Federal grand jury and question her about the source from whom Sanders had received the Calverton hangar evidence.[11] Caving to pressure from the Feds, CBS returned the sample and never aired an interview that Borjesson had taped with Sanders. She didn't have to face the grand jury, but a few weeks later, she says, she was fired.

So far, as this book goes to press, my evidence is still intact. The only formal contact I've had with the FBI is in my letter to Caproni requesting an interview. I have sent another letter to Warden Hood appealing his decision and asking him for any evidence he might have that I would constitute a security risk. And after he piqued my interest, I added an additional request: that I be allowed to talk to the most notorious inmate in the subterranean prison: Scarpa's former cellblock "buddy," Ramzi Ahmed Yousef.

If the 9/11 Commission had relied on the word of Yousef's uncle Khalid Shaikh Mohammed in pushing the date for the genesis of 9/11 from 1994 to 1996, I wondered what secrets the Mozart of Terror might hold if I confronted him face to face.

But it's unlikely that interview will ever be granted. Yousef remains under Special Administrative Measures (SAMs), which keep him in solitary and prevent him from meeting with anyone from the outside world, other than his attorneys. Though he remains a treasure trove of information about al Qaeda, and spoke freely to FBI agents and federal prosecutors after his capture, once he was deposited in a seven-by-twelve-foot cell at the Supermax, Yousef was forever silenced. Today, except for a crude attempt by local FBI agents to shake him down on the morning of 9/11, the true mastermind of the 9/11 attacks remains an enigma.

CAST OF CHARACTERS AND MAJOR EVENTS

THE 9/11 WIDOWS

The "Jersey Girls": (from left) Patty Casazza, Lorie van Auken, Mindy Kleinberg, Kristen Breitweiser

Beverly Eckert

Monica Gabrielle

AL QAEDA 1989–2001

Osama bin Laden

Sheikh Omar Abdel Rahman

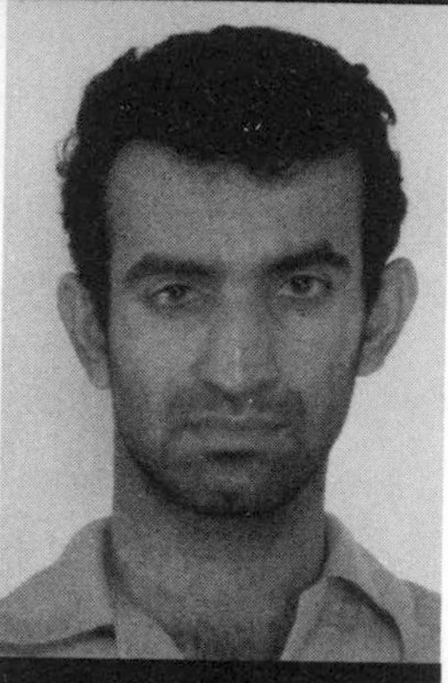

Ramzi Yousef

Dr. Ayman al Zawahiri

Mohammed Atef

Wadih El-Hage

Abdul Hakim Murad

Mahmoud Abouhalima

Mohammad Ajaj

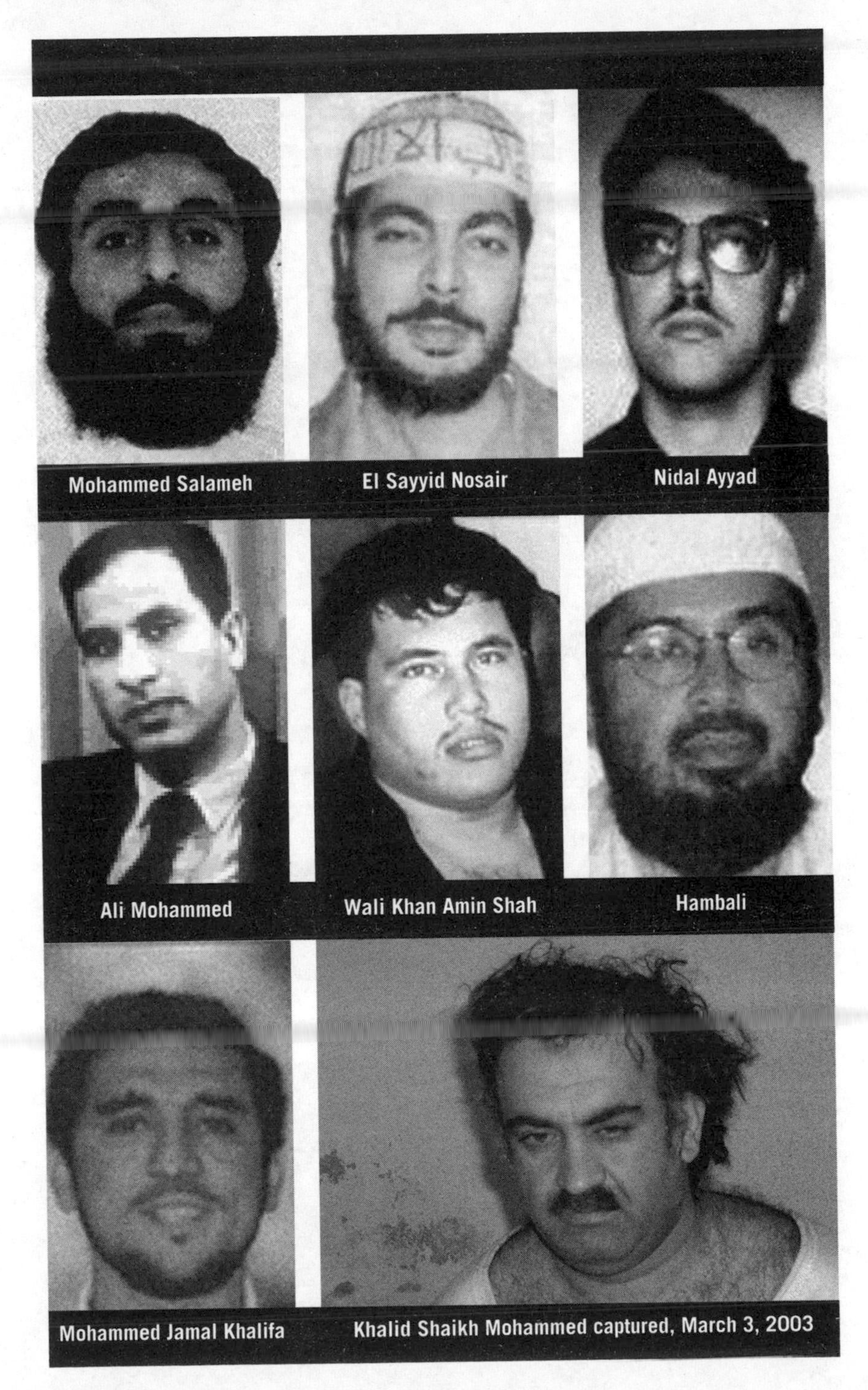

Mohammed Salameh | El Sayyid Nosair | Nidal Ayyad

Ali Mohammed | Wali Khan Amin Shah | Hambali

Mohammed Jamal Khalifa | Khalid Shaikh Mohammed captured, March 3, 2003

THE SCARPA CONNECTION

Greg Scarpa, Sr.

Andrew Goodman, Michael Schwerner, and James Chaney, FBI poster, 1964

Larry Mazza (at right)

R. Lindley DeVecchio

Greg Scarpa, Jr.

Joe Simone

Angela Clemente

THE GOVERNMENT INVESTIGATES

Louis Freeh

Nancy Floyd

Emad Salem

Aida Fariscal

Rodolfo Mendoza

Valerie Caproni

James Kallstrom

THE 9/11 COMMISSION

Gov. Thomas Kean

Rep. Lee Hamilton

Sen. Slade Gorton

John Lehman

Philip Zelikow

Richard Ben-Veniste

Sen. Bob Kerrey

Rep. Tim Roemer

Jamie Gorelick

Sen. Max Cleland

Fred Fielding

Gov. James Thompson

THE BUSH ADMINISTRATION

George W. Bush, 9:06 A.M., 9/11/01

Dick Cheney

Condoleeza Rice

Donald Rumsfeld

Robert S. Mueller

George Tenet

Richard Clarke

THE EVENTS

PAL 434 Bomb damage

TWA 800 Calverton wreckage

9:02 A.M., September 11, 2001

THE EVENTS

September 11, 2001: The Pentagon . . .

. . . and the crash of UA 93, Shanksville, Pennsylvania

Appendix I

COMMISSION ON TERRORIST ATTACKS UPON THE UNITED STATES Testimony of Peter Lance March 15, 2004, at the Commission offices, 26 Federal Plaza, New York, N.Y., before Dietrich Snell, Senior Counsel, and Marco Cordero, Investigator

INTRODUCTION

I am an investigative journalist. A few days after the events of 9/11/01 I began researching the intelligence failures that led up to the attacks. I spent the next 18 months educating myself in an effort to answer two questions: (1) how did the greatest mass murder in U.S. history happen and (2) could it happen again? The result was a book from Harper Collins (ReganBooks) entitled: **1000 YEARS FOR REVENGE: International Terrorism and the FBI--The Untold Story.**

In the course of my research I drew on a number of sources in the U.S. law enforcement and intelligence community. My database of research included dozens of first person interviews, hundreds of pages of declassified documents from the FBI and foreign intelligence services, and more than 40,000 pages of court records and open source material from the print and electronic media. My primary focus was the Federal Bureau of Investigation; specifically the FBI's handling of the original World Trade Center bombing investigation and the hunt for its mastermind Ramzi Ahmed Yousef.

A genius at crafting improvised explosive devices, Yousef fled New York the night of the bombing in 1993 and went on to create the blueprint for the 9/11 attacks. The research uncovered a dotted line between both events that ran like a hot circuit cable from Afghanistan to New York to the Philippines and back to New York. Each dot on that line represented a lost opportunity for the U.S. intelligence community to stop the al Qaeda juggernaut. The evidence I uncovered suggested that while many were culpable, the agency most responsible was the FBI.

As an American first, and a journalist second, I wanted to make my findings available to the 9/11 Commission. Keeping in mind that I was a lone investigative reporter with limited resources and no subpoena power, I was nonetheless able to assemble a trail of evidence dating back 12 years before the attacks. Last fall I contacted the Commission's staff Director Philip Zelikow and offered to testify. On 11/17/03 in an e-mail from Al Felzenberg, the Commission's media contact, my offer was rejected. But several individuals who had been exposed to my book urged the Commissioners to reconsider, including members of the Family Steering Committee and retired FBI Special Agent Joseph F. O'Brien.[1]

Over Christmas of 2003-2004 I sent a copy of my book to Gov. Kean, who responded with the enclosed letter on 1/19/04 saying that he had read my book and referred the matter of my testimony back to Mr. Zelikow.[2] The staff director had earlier contacted me by letter on 1/13/04 and stated that he would put me in touch with Mr. Dietrich (Dieter) Snell, a senior counsel to the Commission.[3]

Mr. Snell contacted me in January of 2004 and after several discussions via phone and e-mail he agreed to my two stipulations: a) that all of my testimony would become a part of the Commission's formal record and that b) as a journalist, I would not be compelled to reveal confidential sources.

With those caveats in mind, I have since pledged to Mr. Snell, following my testimony today, to make available to the Commission on request any documents in support of my findings.

[1] A copy of SA O'Brien's letter to Commission Chairman Tom Kean is attached as **Exhibit A.**

[2] **Exhibit B.**

[3] **Exhibit C.**

My book represented, not only my own enterprise reporting, but a compilation of open source material in the creation of a 12 year TIMELINE leading up to the 9/11 attacks. A written TIMELINE, included in my book, is attached.[4]

The revelations I uncovered raise a number of serious questions that as an American I would hope that the Commission could answer. Thus today, I will make my presentation to the Commission in the form of a series of brief descriptions of my principal findings followed by the questions that I believe the Commission needs to answer in order to fully document the intelligence failures leading up to the attacks.

THE CALVERTON, L.I. SURVEILLANCE

1) FINDING: On four successive weekends in July of 1989, the Special Operations Group (SOG) of the New York FBI office followed a group of ME's (Middle Eastern Men) from the Alkifah Center at the Al Farooq Mosque on Atlantic Avenue in Brooklyn to a shooting range in Calverton, L.I. Of the men under surveillance and subject to dozens of FBI photographs taken over four Sundays, one would be later convicted in the murder of Rabbi Meier Kahane on 11/5/90 (El Sayyid Nosair), three would go on to be convicted in the plot to blow up the World Trade Center on 2/26/93 (Mahmud Abouhalima, Mohammed Salameh and Nidal Ayyad), one would be convicted in the plot to blow up bridges and tunnels around New York City (Clement Rodney Hampton-El), and their leader, an Egyptian named Ali Mohammed, would go on to train Osama bin Laden's personal bodyguard in Khost, Afghanistan in 1990 and undertake the surveillance for the African Embassy bombings in August, 1998.

QUESTIONS:

1) Why did the FBI shut down its August 1989 surveillance of this cell; a group that proved to be tied to al Qaeda and financed directly by Osama bin Laden?

[4] **Exhibit D.**

2) Who in the FBI's New York office was responsible for ending the surveillance?
3) Was anyone in the FBI ever held accountable?

THE KAHANE MURDER & AFTERMATH

2) FINDING: On the night of 11/5/90 in the hours following the murder of Rabbi Kahane, detectives from the NYPD and agents from the FBI searched the home of the shooter, El Sayyid Nosair. In the course of their search they obtained some 47 boxes of evidence suggesting a broad international conspiracy. The evidence seized included 1,400+ rounds of ammunition, books on bomb building, pictures and/or maps of the World Trade Center complex, and audio tapes and notebooks in Arabic in which blind Sheikh Omar Abdel Rahman was reportedly quoted as saying that "the high world buildings" and "edifices of capitalism" should be targets. Further the raid uncovered manuals from the John F. Kennedy Special Warfare Center at Fort Bragg, NC, marked "Top Secret For Training."

Taken into custody at the home were Mahmud Abouhalima and Mohammed Salameh, two of the Calverton shooters. After the NYPD mistakenly labeled the murder a lone gunman shooting, these two men, later convicted in the WTC bombing, were released without charge. Further the evidence shifted back and forth between the NYPD and Federal authorities to the point where much of it was reportedly compromised and rendered inadmissible. The Sheikh's speeches were not translated until after the 1993 WTC bombing.

QUESTIONS:

1) Who was the FBI official responsible for failing to connect the Kahane murder to the Calverton surveillance?
2) Why would the FBI have allowed Abouhalima and Salameh to be released?
3) Did the FBI question them while they were being held by the NYPD as material witnesses on the night of 11/5/90?
4) In the face of this evidence of an international bombing conspiracy, who in the FBI and/or the U.S. Department of Justice made the decision to allow the NYPD to treat the murder as a "lone gunman" incident?

5) Were any FBI or Justice Department officials ever reprimanded for this act of negligence?

THE SHALABI HOMICIDE

3) FINDING: In July, 1990 Sheikh Omar Abdel Rahman, a close associate of Osama bin Laden's who had helped raise funds for the Mujahadeen rebels during the Afghan War against the Soviets, was allowed to enter the U.S. despite his presence on a "watch list." Upon his arrival at Kennedy Airport the blind Sheikh was picked up by Mahmud Abouhalima, then known to the FBI from Calverton and the Kahane murder. Also present was an Egyptian named Mustafa Shalabi who ran the Alkifah Center at the Al Farooq mosque; a hotbed of radical Islamic activity that continued to bring in millions of dollars per year as a conduit for money to the Afghan rebels. The Center was the place from which the FBI's SOG had followed the Calverton shooters in July of 1989.

Soon, despite being housed and endorsed by Shalabi, the Sheikh began a campaign to unseat him at the mosque. Shalabi himself began to fear for his safety. He was preparing to return home to Egypt when, on or about the night of Feb 26, 1991, he was brutally murdered in his home in the Seagate section of Brooklyn, Found shot, stabbed and bludgeoned, Shalabi was identified for NYPD detectives by Mahmud Abouhalima, the very man who had been a material witness in the Kahane murder.

Falsely claiming to be Shalabi's brother, the tall, six-foot-two-inch red-headed Abouhalima was never charged, even though Shalabi was found with two red hairs in his hand. The murder remains unsolved, but from that moment on, Sheikh Abdel Rahman assumed control of the lucrative Alkifah Center at the Al Farooq mosque and Osama bin Laden's al Qaeda had what amounted to a brick and mortar outpost in New York City.

QUESTIONS:

1) Why didn't the FBI link the Shalabi murder to the Calverton-Kahane cell?
2) Why weren't Abouhalima or the Sheikh, for whom the big redhead was acting as a chauffeur, ever charged in the murder, given their motive? With Shalabi's death Rahman inherited a fund raising center that brought in millions of dollars per year.

3) With evidence that Abouhalima was tied to the Calverton Shooters, the Top Secret cache of manuals and bomb making material in Nosair's home as well as the Nosair homicide, why didn't the FBI's New York office connect the dots pointing to evidence of an international conspiracy as far back as February at the time of The Gulf War?

THE FBI'S MOLE INSIDE THE BLIND SHEIKH'S CELL

5) FINDING: In 1991 FBI Special Agent Nancy Floyd recruited one Emad Salem, an Egyptian U.S. naturalized citizen who agreed to work as an FBI asset and successfully infiltrated the cell surrounding Sheikh Abdel Rahman, Abouhalima and Salameh. Risking his life, Mr. Salem was invited up to Attica prison where the then incarcerated Nosair urged the ex-Egyptian army major to engage in a plot to bomb "12 locations" in and around New York City. But shortly thereafter, Carson Dunbar, the former Administrative Special Agent in Charge of the FBI's New York Office, took over the Terrorism Branch.

In his first meeting with Mr. Salem in July of 1992 Mr. Dunbar reportedly changed the terms of Mr. Salem's undercover role, demanded that he wear a wire and be prepared to testify under oath. In addition Mr. Dunbar insisted that Mr. Salem, who had proved his reliability as an asset for months, submit to additional polygraphs to determinate his veracity. As a result, Mr. Salem withdrew, causing the FBI to lose its primary asset inside the blind Sheikh's cell. Once Mr. Salem was out of the picture, the Sheikh contacted Pakistan and Ramzi Ahmed Yousef, a British-trained engineer with roots in the Baluchistan province of Pakistan, came into the U.S.

On 9/1/92, pretending to be an Iraqi refugee, Yousef arrived in New York via a Pakistan International Airlines flight and asked for political asylum. But Martha Morales, an alert INS agent at Kennedy Airport, sought to incarcerate him after her discovery that Yousef's traveling companion was one Mohammed Ajaj, seized moments earlier at JFK carrying fake passports and bomb making paraphernalia.

Agent Morales was overruled by her Supervisor and Yousef slipped into New York City where he quickly linked with Abouhalima, Salameh, and Nidal Ayyad, another shooter from Calverton. All three men were working out of the Alkifah Center, taken over seven months earlier by the blind Sheikh.

Within weeks, in the fall of 1992 Nosair's "12 locations" plot morphed into the World Trade Center bombing conspiracy.

Given three months to wean himself from the Sheikh's cell, Mr. Salem met in the fall of 1992 with Special Agent Floyd who paid him his last $500 fee for working as an FBI asset. At this final meeting Mr. Salem urged her to speak to her superiors. He had picked up word that another plot was afoot. Agent Floyd informed Mr. Salem that she had been frozen out of the terrorism branch and had little influence over her superiors; especially Carson Dunbar.

Mr. Salem exhorted SA Floyd to follow Abouhalima and Salameh and warned that if the Bureau failed to do so, "don't call me when the bombs go off."

My investigation uncovered extraordinary evidence that during the fall of 1992, as Ramzi Yousef built the 1,500 pound urea nitrate fuel oil device that eventually killed six and injured 1000 on 2/26/93, he and his co-conspirators were extraordinarily visible. Mohammed Salameh, who acted as Yousef's bomb building assistant, was involved in no less than three brushes with the local police.

Yousef himself was hospitalized after an accident during which Salameh was driving. The bomb maker, who had made up to $18,000.00 in fraudulent calls on phone credit cards, even ordered chemicals for the bomb from his hospital room. Thousands of dollars in wire transfers came from Germany and the Mideast from bank accounts set up by Salameh and Ayyad and a number of traceable calls were made from a pay phone near Yousef's bomb factory on Pamrapo Avenue in Jersey City to the home of Mahmud Abouhalima who was acting as the chief expeditor for the bomb plot.

Most astonishing, Mohammed Ajaj, Yousef's traveling companion, had made a successful bid in federal court to retrieve his bomb books, and he and Yousef were in regular communication with Ajaj, calling from the U.S. Prison at Otisville, New York, and speaking to Yousef via

Pac Bell's two-way calling system through a hamburger restaurant in Texas. Yet, despite multiple opportunities to apprehend the various members of the bombing cell in the fall of 1992, the evidence suggests that Carson Dunbar refused to approve the surveillance of Abouhalima and Salameh that the FBI's former asset Emad Salem had suggested to Special Agent Nancy Floyd.

<u>If the FBI had approved that surveillance by its elite SOG, Abouhalima and Salameh would have led them directly to Ramzi Yousef in his Pamrapo bomb factory. Yousef would have been captured, the WTC bombing thwarted, and Osama bin Laden's New York al Qaeda cell, which had been fully operational from February, 1991, would have been broken.</u>

<u>Further, because my evidence later revealed that Yousef was the British-trained engineer and architect of the 9/11 plots (which he set in motion in Manila in the fall of 1994), Yousef's capture would have interdicted or prevented the ultimate attacks of 9/11.</u>

QUESTIONS:

1) Why didn't the FBI take seriously Mr. Salem's suggestion to follow Abouhalima and Salameh?
2) Why didn't Carson Dunbar approve the SOG surveillance? Det. Lou Napoli, a veteran member of the NYPD-FBI Joint Terrorist Task Force, told me in an interview that he and FBI Special Agent John Anticev (who had tracked Abouhalima for years) failed to learn of the big Egyptian's whereabouts in the fall of 1992 because "he went to Jersey." But it defies belief that the FBI would have been thwarted by Abouhalima's move from Brooklyn to an apartment in Woodbridge, New Jersey, where he lived openly with his German wife and four children adjacent to the apartment of his brother Mohammed, another six-foot-two-inch redhead. Thus,
3) Was the FBI's failure to track Abouhalima and apprehend Yousef a result of Mr. Dunbar's mistrust of SA Floyd or her asset Emad Salem--a fact that Dunbar admitted under oath in the Landmarks trial?
4) Was anyone in the FBI ever sanctioned for this failure to heed undercover asset Salem's advice and follow Abouhalima and Salameh, who were interacting with Yousef daily as he built the WTC bomb?

THE FAILURE OF CONGRESSIONAL OVERSIGHT

6) FINDING: In a memo published as an appendix to my book,[5] Staff members of the U.S. Senate Judiciary Committee wrote to Sen. Orrin G. Hatch, Republican of Utah, on 12/5/95 urging him to investigate "the FBI's involvement before, during and after" terrorist incidents--specifically citing the Bureau's handling of the WTC bombing investigation. They noted in a memo footnote, "We have information that some instances, like the World Trade Center, could have been prevented if the relevant agencies had worked in concert with each other in the sharing of information . . . If they had shared with each other, there is at least a strong possibility that they would have identified the World Trade Center as a target before the bombing."

Sen. Hatch reportedly refused to heed his staff's recommendations and conduct that hearing, which would have shed important light on the FBI's failures at a critical time. As we will see later, by the spring of 1995, the Justice Department had probative evidence from the Philippines National Police that Ramzi Yousef and his uncle Khalid Shaikh Mohammed had set in motion the plot that culminated on 9/11.

QUESTION:

1) If Congressional overseers of the Bureau like Sen. Hatch had examined Yousef's first date with the World Trade Center, they might well have uncovered details of the plot that led to his second. Why weren't these hearings held?

THE VINDICATION OF AGENT FLOYD'S ASSET

7) FINDING: Within hours of the 2/26/93 WTC bombing, the newly appointed U.S. Attorney for the Southern District of New York, Mary Jo White, ordered Det. Napoli and Agent Anticev to quickly bring Emad Salem back into the fold. Within days he had given the Bureau sufficient information to apprehend the fugitive Mahmud Abouhalima in Egypt, and within weeks Mr. Salem bravely agreed to go back undercover to infiltrate the blind Sheikh's cell.

[5] **Exhibit E** Memo to Orrin G. Hatch from Sen. Judiciary Committee staff members 12/5/95 reproduced on pp. 519–521 of *1000 Years for Revenge.*

By the late Spring of 1993 Mr. Salem had worked with the FBI to set up a sting operation that led to the arrest and conviction of the blind Sheikh and 9 others in the Landmarks (or Day of Terror) plot to blow up, among other targets, the United Nations building, the Lincoln Tunnel, the George Washington Bridge and FBI Headquarters at 26 Federal Plaza.

Mr. Salem went on to become the Government's linchpin witness in the Landmarks Trial and was ultimately paid more than $1.5 million for doing what he could have done for the FBI in the fall of 1992 for $500 a week. Mr. Salem proved to be arguably the most valuable undercover operative the FBI had to that date in the incipient war on terror, yet rather than being rewarded, Special Agent Nancy Floyd, who had recruited and nurtured him as an asset, became the subject of a five-and-one-half-year investigation by the Bureau's Office of Professional Responsibility (OPR).

During this period SA Floyd was subjected to isolation by her peers and humiliation in the tabloid press, when false information was leaked to the New York Post suggesting that she had been having a romantic affair with Mr. Salem. Further she was prevented from advancing in the FBI or transferring to her Office of Preference (OP).

Ultimately SA Floyd was suspended for two weeks--forced to surrender her badge and her gun and put on the street--for her alleged "insubordination" to Carson Dunbar, the ASAC who had failed to heed her advice and that of Emad Salem and authorize the surveillance of Abouhalima and Salameh--key Yousef cohorts--prior to the 1993 bombing.

For his actions, Mr. Dunbar received the extraordinary reward of being allowed to maintain his federal pension by transferring to the ATF in Washington during the same period that he was working fulltime as the Superintendent of the New Jersey State Police.

QUESTION:

1) Why was agent Floyd the ONLY Federal employee known to have been sanctioned for the FBI's failures in the WTC bombing conspiracy, when she was the one agent in the best position to have stopped the 2/26/93 incident?
2) Why was Carson Dunbar given this special retirement deal when the evidence uncovered suggests that he

was more responsible than any single member of "management" in the FBI's New York office for the Bureau failures leading up to the 1993 WTC bombing?

THE BUREAU LOSES A KEY CO-CONSPIRATOR

8) FINDING: Within days of the bombing, after Mohammed Salameh had been arrested by Bureau agents outside DIB Leasing, the Ryder truck dealership in Jersey City, the FBI's Newark office located one Abdul Rahman Yasin, a U.S.-born Iraqi who had been a principal member of Yousef's bombing cell.

But after convincing FBI agents that he knew nothing, they let Yasin go. The next night he was in Baghdad and subsequently Federal agents learned that Yasin had been a key member of Yousef's cell. As of the U.S. invasion of Iraq in March, 2002 the price on Yasin's head was $25 million.

QUESTIONS:

1) What members of the FBI Newark office were responsible for Yasin's release?
2) Have any agents been held responsible? If not, why?

YOUSEF PLANS THE 9/11 ATTACKS AS EARLY AS 1994

9) FINDING: Following his flight from New York on the night of the WTC bombing, Ramzi Yousef made two attempts on the life of the Pakistani Prime Minister and reportedly set a bomb that killed 26 Shiites at a shrine in Iran. By the fall of 1994 he had set up a cell in the Philippines with his uncle Khalid Sheikh Mohammed (KSM), Abdul Hakim Murad, a lifelong friend who had trained previously in four U.S. flight schools, and Wali Khan Amin Shah, an Uzbeki veteran of the Afghan campaign who Osama bin Laden called "The Lion," for his reputed bravery in the war against the Soviets.

By November of 1994 Yousef had established a bomb factory in the Dona Josefa Apartments in Manila. With the support of the Abu Sayyaf Group (ASG), an al Qaeda-associated cell, and funded via a Malaysian cutout company called Konsonjaya--bankrolled by bin Laden's brother-in-law, Mohammed Jamal Khalifa--Yousef, KSM, and the other two principal cell members were planning three operations:

First, they intended to kill Pope John Paul II, who was traveling to Manila on 1/12/95. Second, they plotted to plant small "bomb triggers" powered by Casio DBC-61 watches on 11 U.S. bound commercial airline flights from Asia. Assembled from apparently innocuous components that they would smuggle on board the two-leg flights, Yousef, Murad, KSM and Shah planned to assemble the devices and plant them in the life jacket pouches along the 25 rows of the Boeing 747's.

Yousef himself had conducted what he called "a wet test" of a Philippines Airlines 747 on 12/11/94 and, but for a miscalculation on placement (the bomb was in the 26th row just shy of the center fuel tank) the flight with 273 passengers and crew would have been blown out of the sky. Instead, a Japanese national in seat 26K, where Yousef had planted the bomb, was killed and PAL Flight #434 made an emergency landing in Okinawa.

This second plot was dubbed "Bojinka" by Yousef after the Serbo-Croatian word for "big noise."

The third plot, which Yousef conceived during bomb training with Murad in Lahore, Pakistan in September of 1994, was the hijack-airliners-fly-them-into-buildings scenario that culminated on Sept. 11, 2001. This plot was separate and distinct from Bojinka and Yousef's cohort Murad, a commercial pilot trained at flight schools in the UAR, the Philippines, and Texas, North Carolina, New York and California, was to be the lead pilot. Earlier, during his training in 1992, Murad had surveilled the World Trade Center, and he vowed along with Yousef that he would return to take down the Twin Towers by striking them from above with fuel-laden commercial aircraft.

All three of Yousef's plots might have succeeded but for an accidental fire in Room #603 of the Dona Josefa on the night of January 6, 1995 and the heroic intervention of PNP Police Capt. Aida Fariscal. Immediately following the incident, Murad was arrested by PNP officers and taken to Camp Crame, home of the Philippines Presidential Security Group in Quezon City, Manila. There, over 67 days he was subject to harsh interrogation.

Yousef and KSM escaped followed the fire, but Yousef was apprehended in Islamabad, Pakistan, a month later. Though the FBI falsely took credit for his capture, affected after a tip to the U.S. State Dept. via its Rewards for Justice Program, FBI agents failed to search the 20-room Su Casa Guesthouse, a bin Laden controlled boarding house where Yousef was staying. As such, they failed to capture Khalid Shaikh Mohammed, who was not only staying in a room on the ground floor, but actually gave an interview to TIME magazine in his own name.

Wali Khan Amin Shah was captured days after the Dona Josefa fire, but escaped from Camp Crame, only to be apprehended the following year in Malaysian and rendered back to the U.S. for trial. Abdul Hakim Murad was also extradited back, but not before confessing to PNP Col. Rodolfo Mendoza precise details of the 9/11 plot, including the fact that Yousef and KSM (as early as 1994) had chosen six targets including the WTC, CIA headquarters, the Pentagon, a nuclear facility, and the Sears and Transamerica Towers. They also had as many as 10 Islamic radicals then training for the operation in U.S. flight schools. Col. Mendoza gave all of the details of this third plot to the U.S. Embassy in Manila, and my investigation revealed that the information was received by the Justice Department because it was referenced in a series of FBI NO/FORN memos in 1995.

In 1996 Yousef, Murad, and Shah were tried in U.S. Court for the Southern District of New York. KSM, though indicted with Yousef in 1996, remained a fugitive, and his part of the indictment remained sealed until 1998.

In 1996, the Feds appeared so desperate to trap Yousef that they set up a "pass through" system on his Nine South Tier of the Metropolitan Correctional Center, the federal jail in Lower Manhattan. Seeking the cooperation of an accused member of the Columbo Crime Family named Gregory Scarpa, the FBI supplied Scarpa with a phone allowing him to make outside calls, in the hope that he would induce Yousef to contact al Qaeda associates outside of prison and that the FBI might monitor the calls to gather intelligence.

But Yousef apparently turned the tables on the Feds and was able to make a number of untraced outside calls, including at least one to his uncle Khalid Shaikh Mohammed, who, by 1996, was in the early stages of perfecting the plot that culminated on 9/11.

With the weight of the evidence against Yousef and a strong presentation by Assistant U.S. Attorneys, the bomb maker and his co-conspirators were convicted in the Bojinka plot.

But in the 6,000-page Bojinka trial transcript there isn't any mention of the third plot that became the 9/11 attacks, even though Col. Mendoza had delivered the details to Federal authorities in the spring of 1995. Further, though he was the chief interrogator of Murad, Col. Mendoza was never called to testify at the trial in the S.D.N.Y., and his co-interrogator Major Alberto Ferro of the PNP, who was with Col. Mendoza for most of Murad's questioning, apparently lied under oath when he stated that he did not recall the identities of the others who had interrogated Murad.

FBI agent Frank Pellegrino, who had traveled to Manila in mid-January, 1995, was present at Camp Crame during almost two months of Murad's interrogation and was co-author of the FBI 302 memo describing the airborne interrogation of Murad, but SA Pellegrino testified at the Bojinka trial that he was unaware that Murad was being interrogated by the PNP until mid March of 1995, two months after his arrival at Camp Crame--a statement that defies belief given the close cooperation between U.S. and PNP authorities in the extradition of Murad.

Further, an FBI 302 memo[6] in which Agent Pellegrino memorialized the questioning of Murad during his rendition to the U.S. quoted Murad as advising "that RAMZI wanted to return to the U.S. in the future to bomb the World Trade Center a second time." Yet no reference to Yousef's third plot was ever introduced at the Bojinka trial.

And though the hunt for Yousef was very public, resulting in the reproduction of his picture on thousands of wanted posters and matchbooks worldwide, the identity of his uncle and 9/11 co-conspirator KSM remained secret. Federal officials didn't reveal the presence of KSM as a Bojinka conspirator until January 1998, and he wasn't placed on the FBI's list of terrorists until October 2001, after 9/11, though it is clear now that KSM executed the plot conceived by his nephew Ramzi; a plot known to the FBI by May of 1995.

In an interview with me, the most detailed he has given to any journalist to date, Col. Mendoza was so certain about the details of the 9/11 plot which Murad had confessed had been set in motion in 1995, that on the

[6] **Exhibit F** Reproduced as pp. 499–516 in *1000 Years for Revenge.*

evening of Sept. 11, 2001 (Philippines time), after witnessing the collapse of the WTC's South Tower, the first words out his mouth were, "They have done it. They have DONE it!"

Col. Mendoza was certain that the Yousef-Khalid Shaikh Mohammed al Qaeda cell had finally carried out the plot he had warned the U.S. Feds about in the spring of 1995.

QUESTIONS:

1) Why were the details of Col. Mendoza's revelations about this third plot--separate from Bojinka and the plot to kill the Pope--apparently ignored by officials in the FBI and the Justice Department?
2) What officials in the FBI or DOJ gave the command to limit the prosecution of Yousef to the Bojinka plot and the original WTC bombing conspiracy, effectively excluding this probative evidence of the plot that culminated on 9/11?
3) Why was Khalid Shaikh Mohammed treated differently as a fugitive than his nephew Ramzi? Why wasn't a public hue and cry issued for him and a public reward offered in the manner that successfully affected Yousef's capture?
4) Why wasn't Col. Mendoza called to testify in the Bojinka case and why were PNP officer Ferro and FBI agent Pellegrino allowed to give misleading answers to the jury that would have prevented the media covering the trial to learn the details of Yousef's third plot?

THE PHOENIX CONNECTION

10) FINDING: Harry Ellen, a U.S. citizen and convert to Islam who had worked as an FBI asset via Squad Five of the Phoenix office in the mid 1990s, told me that in the fall of 1996 he had told his control agent that a suspicious-looking Algerian pilot had met with a man who had bragged to him (Ellen) about being an associate of Ramzi Yousef. Ellen told me that after pressing Special Agent Ken Williams and suggesting that he get to know the pilot, he (Ellen) had been told by SA Williams to "leave it alone."

Following the Sept. 11 attacks, Ellen recognized the Algerian pilot as one Lotfi Raissi, who was arrested by

British officials and indicted by the U.S. Justice Dept. on charges of fraud and giving false information on his application for a pilot's license. A British judge later set Raissi free, concluding that there was insufficient evidence to tie him directly to the 9/11 conspiracy, but the FBI found evidence that on three occasions Raissi had been in close proximity to Hani Hanjour, believed to be the pilot of AA Flight #77 that struck the Pentagon.

SA Williams went on to be celebrated following 9/11 as the author of the "Phoenix memo" sent to the FBI HQ on 7/10/01, in which he suggested that the Bureau monitor U.S. flight schools for suspicious Islamic pilots. He had operated Ellen as a successful asset for several years, and Ellen had brokered a deal with five of the most virulent Palestinian terrorist organizations to allow the safe passage of a series of doctors into Gaza to bring medical aid to the refugee camps.

Then in 1998 Williams had a falling-out with Ellen, who accused the FBI agent of threatening to blow his cover with the Palestinians, revealing him to be an FBI asset. Ellen and Mark Flatten, a reporter for the East Valley Tribune, a Phoenix area newspaper, told me that they believed the timing of Williams' Phoenix memo had come more than four and a half years after Ellen first warned him of Raissi and another man tied to Yousef, because Williams wanted to cover himself before an embarrassing series of articles appeared in the Tribune in which his break with Ellen was to be recounted by Flatten.

In May of 2002, in testimony before Congress, SA Williams said that if FBI officials had acted on his recommendations in the Phoenix Memo he believed the Sept. 11 attacks might have been thwarted.

QUESTIONS:

1) Given the details Murad confessed to Col. Mendoza in 1995 about the presence of Islamic radicals tied to Yousef then training in U.S. flight schools, and given Harry Ellen's reliability as an FBI asset who had successfully brokered a deal to deliver medical aid to Gaza, why did it take SA Ken Williams more than four and a half years to act on Ellen's suggestion that he monitor Islamic pilots?
2) Did Williams send any other communiqués to FBI Headquarters in the years leading up to July 2001,

in which he suggested that Islamic pilots should be monitored?

3) During the mid-1990s, when Ellen first warned SA Williams about suspicious Islamic pilots in Arizona, SAC Lupe Gonzalez admitted that terrorism was "job four" in the Phoenix FBI office behind "organized crime and drugs, white collar crime, and crime on Indian reservations." Yet he had been preceded as SAC in Phoenix by Bruce Gebhardt who went on to become No. 2 at the FBI under Director Robert S. Mueller, and SAC Gonzalez was later promoted to head of the Dallas FBI office. Were any agents or SACs sanctioned for the failure of the FBI's Phoenix office to detect the presence of Islamic pilots who were training as early as the mid-1990s?
4) Like Nancy Floyd, whose career in the FBI was permanently damaged after she attempted to thwart Yousef's New York cell, Harry Ellen claims that his life has been ruined by the FBI following his falling out with SA Williams. Has the FBI taken any specific reprisals vs. Mr. Ellen? Why hasn't he been called as a witness before the 9/11 Commission?
5) Why hasn't Special Agent Williams been called to testify under oath as to the precise time he first became aware of the threat from Islamic radicals training in Arizona area flight schools?

AN EGYPTIAN MOLE IN THE FDNY?

11) FINDINGS: Fire Marshal Ronnie Bucca, who perished on the 78th floor of the South Tower on Sept. 11, was working as a terrorism liaison to the FDNY at Metrotech, the fire department's headquarters, in 1998 when he uncovered evidence that Ahmed Amin Refai, an Egyptian naturalized citizen working as an accountant for the FDNY, had obtained the blueprints of the World Trade Center from the FDNY prior to the 1993 bombing. Further, Fire Marshal Bucca discovered that Mr. Refai was an intimate of blind Sheikh Omar Abdel Rahman and had been photographed on the Sheikh's arm acting as his bodyguard in the months prior to the WTC bombing.

Further, Fire Marshal Bucca learned that Mr. Refai had been questioned twice by Federal agents in 1994, who had apparently failed to follow up on his activities. Then in 1998 Fire Marshal Bucca learned that Mr. Refai had told multiple lies to obtain a second I.D. card allowing entrance to Metrotech the FDNY's SECURE headquarters which contained blueprints of most of the city's major buildings including the updated plans of the WTC complex. Mr. Bucca presented his findings to the FBI's JTTF and they were ignored.

Following his death on Sept. 11, Fire Marshal Bucca's widow, Eve, contacted the FDNY, who sent the Refai file to the FBI a second time and a second time it was ignored. In mid-October of 1995 after I sent a copy of the Refai file to an official in the Department of Homeland Security, the FBI opened an investigation into Mr. Refai. In mid-July 2003, in an interview with me for my book, Mr. Refai denied that he had obtained the WTC blueprints in 1992 but admitted that he had been questioned by Federal agents. Shortly thereafter he left his home in Middletown, New Jersey, and has not been seen by neighbors since.

QUESTIONS:

1) Could Mr. Refai have been an al Qaeda mole working inside the FDNY?
2) Why did Federal agents reject probative evidence in 1994 that Mr. Refai had obtained the plans of the WTC prior to the 1993 bombing?
3) Why did the Joint Terrorist Task Force spurn Fire Marshal Bucca's urgent request that the Refai case be examined following evidence that Refai had lied to obtain a second I.D. to Metrotech in 1998?
4) Why did the FBI spurn the Refai investigation a second time following 9/11?
5) What is the current status of the FBI's Refai investigation?

THE ROAD TO 9/11

In summation, my investigation of the 12 years leading up to the 9/11 attacks shows dozens and dozens of blunders by experienced FBI agents, particularly agents in the New York office (the bin Laden office of origin) and the Phoenix office of the FBI. The negligence crosses three presidential administrations and culpability can be apportioned almost equally across all three.

It was on the watch of President George Herbert Walker Bush that the FBI was in the best position to stop Ramzi

Yousef as he built the first WTC device. But due to apparent ignorance of the threat posed by OBL's incipient al Qaeda cell and the intransigence of FBI officials like Carson Dunbar, the Bureau failed to prevent the first WTC attack in 1993.

Having lost Yousef the first time in New York, the Bureau and the Justice Department compounded the problem during the Administration of President Bill Clinton by narrowing the prosecution of Yousef after his capture, ignoring evidence of the third plot set in motion by Yousef, failing to capture his uncle KSM and failing to investigate the leads pointing to the 9/11 attacks supplied by Col. Rodolfo Mendoza of the PNP.

Further, during Clinton's presidency Mohammed Jamal Khalifa, OBL's brother-in-law, was in U.S. custody on his way to Manila in December of 1994 when the State Department pressured the DOJ to release him back to Jordan for trial. Rather than holding onto this key asset, who was reportedly the chief financial officer of al Qaeda, the State Dept. forced his extradition to Jordan, where a witness in his murder trial recanted and Khalifa's conviction was overturned.

This represented the next major lost opportunity to fully appreciate the ongoing plans of OBL's al Qaeda terror network as KSM perfected the 9/11 plot.

Finally, the last chapters of my book recount the staggering amount of intelligence of an impending attack revealed in the months of 2001 to officials of the second Bush Administration.

The intelligence spikes suggesting an attack by airliners were so numerous by July 6, 2001, that Richard Clarke, the terrorism "czar" under Presidents Clinton and George W. Bush, called a White House meeting with officials from the FBI, FAA, Coast Guard, Secret Services, and INS, warning that "something really spectacular is going to happen here and it's going to happen soon." The next day Clarke chaired a meeting of the National Security Council's Counterterrorism Security Group (CSG) and ordered a suspension of all non-essential travel by the staff.

On July 19 Attorney General Ashcroft left on a trip to his home state of Missouri aboard a private jet. When asked by reporters why he didn't use a commercial jet as was customary, the DOJ cited what it called a "threat assessment" by the FBI. The A.G., they said, had been advised to travel only by private jet for the rest of his term.

My book recounted in detail the role of Egyptian radicals surrounding Osama bin Laden. Contrary to the testimony before this Commission by Laurie Mylroie, author of the disputed theory that Ramzi Yousef had been an Iraqi agent, my investigation found ZERO EVIDENCE of any ties between Saddam Hussein and OBL in the 9/11 attacks.

In fact, the foreign nationals that helped bin Laden seize control of the MAK fund-raising network (which became al Qaeda) and served at the top of al Qaeda's power structure were all Egyptian radicals; from Mohammed Atef to Dr. Ayman al-Zawahiri and blind Sheikh Omar Abdel Rahman, bin Laden's war against the U.S. has been directed and sponsored largely by Egyptians. In particular, blind Sheikh Rahman has had a central role in both attacks on the World Trade Center, first as the spiritual leader of Yousef's 1992 cell and secondly as a primary motivating factor in al Qaeda's later attack.

In the weeks before the bombing of the U.S.S. Cole in October of 2000 bin Laden issued the now-famous video fatwa in which he appeared with the blind Sheikh's son and admonished the U.S. to free Abdel Rahman, then in U.S. prison.

The controversial August 6, 2001, Crawford Texas PDB reportedly contains a reference to a British intelligence report about a plot to hijack a plane to free the blind Sheikh and just five days prior to the 9/11 attacks the Taliban offered to exchange the eight Christian aid workers being held in Kabul for Omar Abdel Rahman.

This makes the May 17, 2002, statement by National Security Advisor Condoleezza Rice all the more astonishing. In a press conference she said, "I don't think anybody could have predicted that these people would take an airplane and slam it into the World Trade Center, take another one and slam it into the Pentagon; that they would try and use an airplane as a missile."[7]

Given my findings that Col. Mendoza had relayed details of that very scenario to the U.S. Embassy in Manila in 1995, Dr. Rice's statement makes her appear either to be

[7] CNN, 5/17/02.

a liar or an incompetent, and, as an American citizen, I don't know which would condition would be worse in a National Security Advisor. <u>That is why I urge this Commission to subpoena Dr. Rice as a witness and compel her to testify under oath.</u>

In the decades ahead NO ONE will ever recall what jobs the ten 9/11 Commissioners held prior to their appointment to this crucial body. But if they allow themselves to be limited in the information they gather or if they allow evidence to be cherry-picked to fit a specific political agenda, the America public will hold each and every one of them accountable along with the Commission staff.

As I said at the outset, the attacks of Sept. 11, 2001 represent the greatest unsolved mass murder in U.S. history. It is incumbent upon this Commission to solve that crime and give the American public a full, complete, and honest rendering of the intelligence failures that led up to it. If the Commission fails to do so, that continuing threat from al Qaeda will only put this country at further risk.

I thank you for this opportunity to present an overview of my findings and stand ready and willing to back up my testimony today with documentary evidence.

Respectfully submitted
by Peter Anthony Lance
3/15/04 New York City

EXHIBIT A

Joe O'Brien Investigations, Inc.
67 Wall Street – 22nd Fl. - New York, NY 10005-3101
(800) 670-2347

Gov. Thomas Kean, President
Drew University - Mead Hall
36 Madison Avenue
Madison, N.J. 07940

Dear Gov. Kean.

As a former agent who served 19 years in the FBI (mostly in New York City) and as the proud recipient of the U.S. Attorney General's Distinguished Service Award for my investigative work against Organized Crime, I am acutely aware of the importance of your mission as Chairman of the 9/11 Commission.

Last summer I became aware of the work being done to uncover the truth behind the Bureau's intel failures leading up to the attacks by an investigative reporter named Peter Lance. I had never met Mr. Lance, but after we connected I was impressed with the work he had done and offered to help him in any way I could.

In early July I accompanied Mr. Lance to the home of one Ahmed Amin Refai, an Egyptian naturalized citizen who had retired as an accountant from the New York City Fire Dept. I had examined Mr. Lance's investigative files on this individual and seen a video and still picture of him cheek to jowl with the blind Sheikh, one of the most dangerous terrorists ever admitted to our country.

Mr. Lance, it seemed to me, had developed probative evidence, originally unearthed by Fire Marshal Ronnie Bucca who died on 9/11, that Mr. Refai was acting on behalf of the Sheikh's cell when he obtained the blueprints of the Twin Towers before the bombing by Ramzi Yousef in 1993. Sensing that Mr. Lance might be in some danger in interviewing Mr. Refai, I offered to accompany him to Middletown New Jersey after he had learned that Mr. Refai had slipped back into the country from Egypt.

In the course of a 45 minutes interview it was clear to me that Mr. Refai was being evasive, contradictory and, at times, openly untruthful in his efforts to diminish his relationship with the blind Sheikh. During the interview Mr. Lance got Mr. Refai to admit that he had acted as the Sheikh's translator during a crucial INS hearing which could have affected the Sheikh's status in this country.

Mr. Refai also admitted that he was a member of the notorious al-Salaam mosque in Jersey City which had been a focal point for the Ramzi Yousef cell in the months prior to the WTC bombing in 1993. Refai also admitted, after continued questioning by Mr. Lance, that he had been the object of a recent FBI investigation.

Later, after examining some of Mr. Lance's documentary evidence and reading his book cover to cover I was stunned by the depth and breadth of his investigation. For the first time he uncovered detailed evidence of a connection via Ramzi Yousef to the two attacks on the Trade Center in 1993 and 2001 and further, underscored the negligence of certain members in FBI management in failing to stop Yousef prior to the first bombing.

More importantly, with respect to Mr. Refai, Mr. Lance unearthed compelling evidence of an al Qaeda related "sleeper cell" that was active as late as 1999, a time when, we now know Khalid Shaikh Mohammed, Yousef's uncle, was working in Hamburg with Mohammed Atta, another Egyptian, to perfect the 9/11 plot.

Mr. Lance discovered that Fire Marshal Bucca had alerted the FBI's Joint Terrorist Task Force to this evidence on Mr. Refai as early as September of 1999, but he had been spurned – only to perish on the 78th floor of Tower Two on Sept. 11th.

Based on my knowledge of Peter Lance and his investigation I can tell you without a doubt that had Mr. Lance been a Special Agent with the FBI working International Terrorism cases in the 90's, OBL and the al Qaeda terrorist who attacked America would either be in a U.S. prison, a foreign prison or deceased, which is perhaps a more proper sentence for them.

The FBI did not have and still does not have a clue, when it comes to dealing with this kind of criminal element. It took the FBI 30 years to take down the Mafia. This country does not have the luxury of that kind of time, when dealing with these terrorists. The American public deserves some accountability from our government officials for allowing 9/11 to happen. Your Commission holds out the prospect of that. But in order to get to the real truth you need ALL of the facts.

I strongly urge you to read Peter Lance's book and allow him to testify at your upcoming hearings in February. I have studied this matter closely and from my standpoint, 1000 YEARS FOR REVENGE offers the most compelling evidence to date that the attacks of 9/11 could have been prevented.

If you have any questions please feel free to contact me at any time.

Respectfully,

Joseph F. O'Brien
FBI Special Agent (ret.)

EXHIBIT B

DREW UNIVERSITY
Office of the President
Madison, New Jersey 07940
(973) 408-3100

January 19, 2004

Mr. Peter Lance
140 Butterfly Lane
Santa Barbara, CA 93108

Dear Mr. Lance:

Thank you for your recent note and a copy of your book, *1000 Years for Revenge*. I read it with great interest. On behalf of the Commission and its staff, I thank you for sharing your obviously extensive work with us. I know that the staff has read your book and found it helpful in several areas.

I have referred your request to testify to Philip Zelikow, who is setting up the Commission's public hearings.

Again, thank you for sharing your work. Best wishes.

Sincerely,

Thomas H. Kean
President

cc: Philip Zelikow

EXHIBIT C

January 13, 2004

Mr. Peter Lance
140 Butterfly Lane
Santa Barbara, CA 93108

Dear Mr. Lance:

I write to follow up on your correspondence of December 20, 2003, to Chairman Kean regarding your work on the 9/11 attacks. The Commission very much appreciates your interest in appearing as a hearing witness. As I'm sure you can understand, however, given the severe time constraints under which we labor, the Commission's ability to hear public testimony is extremely limited and it, unfortunately, is not possible for us to work you into our crowded hearing schedule.

That said, Commission staff has read your book carefully and is very interested in meeting with you to discuss your work and source materials. Accordingly, I have asked that staff member Dietrich Snell, who heads up the team investigating the 9/11 plot, contact you to arrange such a meeting.

Thank you very much for your interest in assisting our effort.

Sincerely,

Philip Zelikow

Philip Zelikow

TEL (202) 331-4060
FAX (202) 296-5545
www.9-11commission.gov

EXHIBIT D: Timeline can be found in the center section of *1000 Years for Revenge: International Terrorism and the FBI—The Untold Story* (New York: ReganBooks/HarperCollins, 2003).

EXHIBIT E: Memo to Orrin G. Hatch from Sen. Judiciary Committee staff members December 5, 1995, reproduced on pp. 519–521 of *1000 Years for Revenge.*

EXHIBIT F: FBI 302 memo concerning the interrogation of Abdul Hakim Murad, reproduced on pp. 499–516 in *1000 Years for Revenge.*

Appendix II: Documents Relating to the Relationship Between Ramzi Yousef and Gregory Scarpa Jr.

265A-NY-252802

Continuation of FD-302 of GREGORY SCARPA, JR. , On 3/5/96 , Page ___

SCARPA advised that YOUSEF began slipping papers to him, 1/2 sheet of paper rolled up with writing on them. According to SCARPA, YOUSEF writes in sentences. SCARPA advised that when YOUSEF slips him these papers he writes on the paper that he wants them back. SCARPA advised he has only kept the notes for a matter of minutes, approximately 10 minutes, just long enough to write some things down. SCARPA advised he may get one note a day or one every couple of days. According to SCARPA, the number of notes varies according to the circumstances at the mcc. SCARPA advised YOUSEF does not give his notes back to him, but expects SCARPA to return YOUSEF's notes quickly. SCARPA believes YOUSEF throws the notes in the toilet.

SCARPA advised these notes, referred to as "kites" are passed from inmate to inmate, in newspapers, or through holes in the walls of the cells.

SCARPA advised that there is a guard permanently assigned to both ISMAIL and YOUSEF and sometimes they check the newspaper before its given to ISMAIL or YOUSEF and sometimes they do not.

According to SCARPA, YOUSEF told him, "if you're interested, I'll teach you things nobody knows."

YOUSEF told SCARPA I'll teach you how to blow up airplanes, and how to make bombs and then you can get the information to your people (meaning SCARPA's people on the outside). YOUSEF told SCARPA I can show you how to get a bomb on an airplane through a metal detector. YOUSEF told SCARPA he would teach him how to make timing devices.

According to SCARPA, YOUSEF wants to hurt the United States Government and wants to teach SCARPA how. SCARPA advised [illegible] YOUSEF wants to blow things up, but he does not say why.

YOUSEF told SCARPA that during the trial they had a plan to blow up a plane to show that they are serious and then make their demands, or kidnap and hurt a judge or an attorney so a mistrial will be declared. SCARPA advised that blowing up an airplane during the trial seems easy to YOUSEF. YOUSEF never mentioned a specific airline as a target.

According to SCARPA, YOUSEF believes that SCARPA is in touch with people on the outside. YOUSEF told SCARPA if things get going we may be able to hook up, if you're serious my people and your people can meet. SCARPA believes YOUSEF needs help

A. Yousef describes his plan to blow up an airplane in order to effect a mistrial: March 5, 1996.

The Following diagram is an example of Connecting SCR Clock
To The Back of a Digital Clock

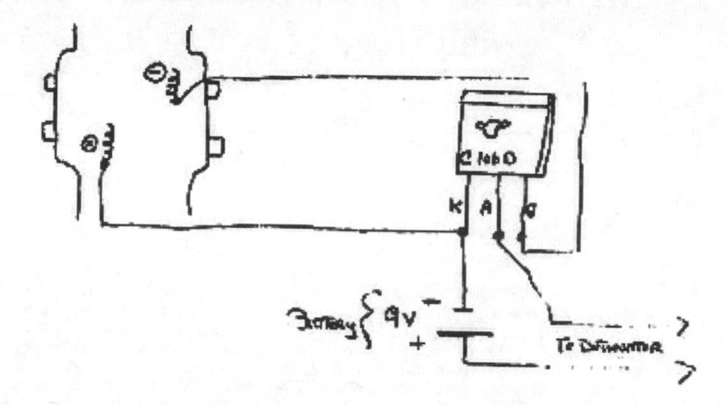

④ To Test The Circuit, Connect a 6 or 9 V Bulb To The Two Wires
going To The Detonator, The Bulb should only operate Whenever The Alarm
Of The Clock Sets off

B. Scarpa's copy of Yousef's bomb-trigger diagram, 1996 (above), and the Casio-nitro device seized from Yousef's Manila bomb factory, 1995 (below).

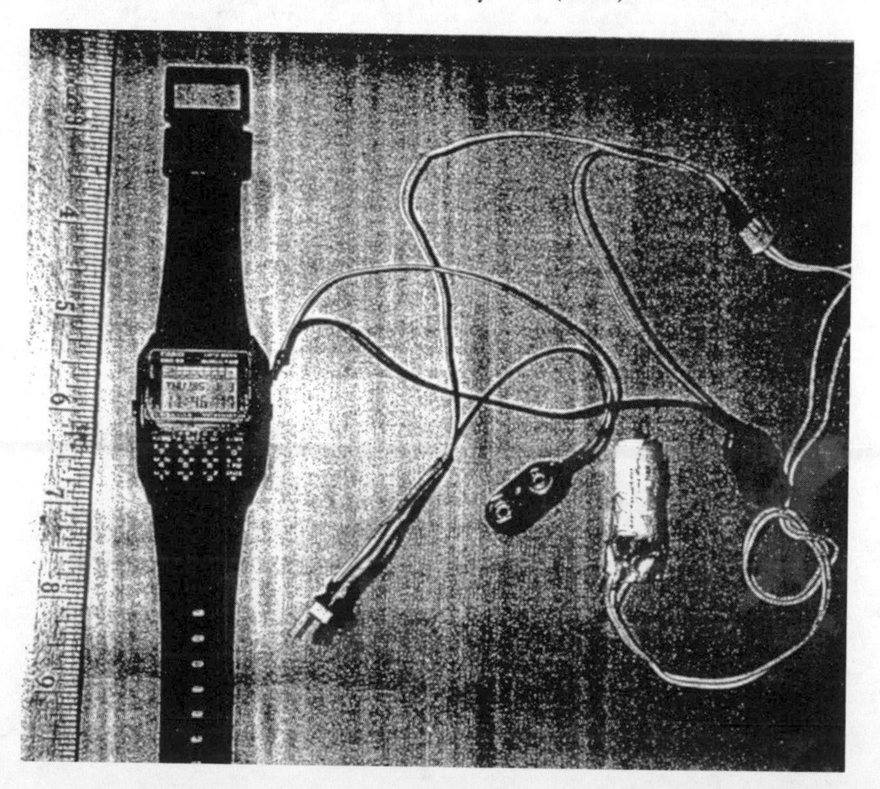

TOP SECRET

OBAID:
SIN > HNG 1- 1234567 (3:45) 0735 - 1120 UA80
744
HNK>SFO 1234567 (12:20) 1300 - 1020 UA806 744
SETTING : 9 -10
TIMER: 10 HR
BOJINGA 19-20 NRT DATE 4
HNG>SIN 2- 1234567 (3:35) 2005 - 2340 UA80
744
SIN>HNG 1234567 0735 - 1120 UA806 744
HNG>SOF (12:20) 1300 - 1020 UA806 744
SETTING 9:30 PM-10:30 PM
TIMER: 23 HR
BOJINGA 20:30-21:30 NRT DATE 5
HOME:

5	0055 - 0410	PK 77	SIN - KAC
5	0040 - 0530	EK077	SIN - DXB
5	0700 - 0950	MS 870	DXB-KAC
5	0815 - 1105	EK 600	DXB-KAC

C. Yousef's Bojinka schedule, as found on his Manila laptop (above). Scarpa's note to Yousef regarding Roma Corporation (below).

May 9

How are you Buddy?
I Hope your CASE Went Well For you ToDay.
Listen Bo as I Told you My Lawyer is Pretty Good.
As Soon as He Received The INFO From my people, He
Brought It right up.

[redacted] Phone- AS I already gave you
[redacted] FAX
You're To Say; Ronnie Calling - I'd Like To make a phone
call;
you Send your FAX To;
Geoge Smith
c/o Roma Corp
[redacted] Suite 2252
N.Y. N.Y. 10010

How To Smuggle Explosives INTo an airplane. SUN 5/19

1. When Talking about Smuggling Explosives into an Airplane, It is meant By That Explosive Substances which are going To Be placed INTo an airplane Later, To Be Used For Blowing up The airplanes. There Fore The quantity of Explosive Substaces Which We'll Be Talking about, is The quantity needed To Blow-up an airplane, Which is equivelant To 300gM of TNT For 747-400 airplanes, aT 30,000 Feet altitude. For Substanses Which are more powerful Than T.N.T., The quantity needed Would Be Less Than 300 gm depending ON it's power Compared To T.N.T.

2. All metalic substances, or substances which contains metals, CANNOT Be Used or Smuggled INTo The airplane Because They are easily detected By X-Ray machines and Metal Detectors, There Fore, all Azideas and mercury compounds Explosive Substances Should Not Be Used.

3. all Explosive Substances of A Density higher Than 2 Kg/L Should NoT Be Used due To The possibility of Detecting Them by X-Ray Machines.

The Following Explosive Substances Have a Density Less Than 2 gm can Easily Be Smuggled.

1. Tetrazene (Guanyl Nitrosoaminoguanyltetrazene)
2. Acetone Peroxide
3. R.D.X.
4. H.M.T.D. (Hexamethylenetriperoxidediamine)
5. D.D.N.P. (Diazodinitrophenol)
6. H.M.X. (cyclotetramethylenetetranitramine)

All Liquid Explosives CAN Be Used

1. When an Explosive Substance is smuggled INTo AN Airplane, Then it CAN Be assembled easily INSide The airplane
2. A ~~Det~~ DeTONATor CAN Be HID in a Heel of a Shoe
3. The Wiring and 9V Battery CAN Be HIDDEN INSide a Shaving machine or a Toy.

The above Explosive Substances which are IN Powder Form can also Be Smuggled IN AN Airplane easily By Hiding Them in The Holes in The Heal of a Shoe, OR By putting Them IN Medicine Capsules.

D. Scarpa's copy of Yousef's May 19, 1996, kite. Note the mention of RDX—the very explosive found on TWA 800 after its crash two months later.

Forensic Intelligence International, LLC
***the Kauth house*, 318 Cooper Avenue, Hancock, Michigan 49930; 906-370-9993**
706-294-9993 (mobile), 603-452-8208 (fax & voice mail), sdresch@forensic-intelligence.org

Monday, April 5, 2004

Urgent & Confidential

Thomas H. Kean, Chairman
Lee H. Hamilton, Vice Chairman
National Commission on Terrorist Attacks Upon the United States
301 7th Street, SW
Room 5125
Washington, DC 20407

Re: Pre-September 2001 FBI and Dept. of Justice Intelligence on Terrorism (R. Yousef)

Dear Chairman Kean and Vice Chairman Hamilton:

Our monitoring of the Commission's hearings indicates that significant intelligence concerning the terrorist threat to the United States obtained between 1996 and early 2001 by various components of the U.S. Department of Justice has **not** been provided to the Commission.

This intelligence was provided to agents of the New York office of the FBI and senior prosecutors with the offices of the U.S. Attorneys for the Southern and Eastern Districts of New York by an informant, Gregory Scarpa, Jr., who developed a close personal relationship with Ramzi Ahmed Yousef while Yousef and Scarpa were incarcerated at the Metropolitan Correctional Center (MCC) in New York in 1996 and 1997, with supplemental information provided in 2000 by Scarpa to an official of the U.S. Bureau of Prisons after he was transferred to the Administrative Maximum Security (ADMAX) Penitentiary in Florence Colorado, to which Yousef had already been transferred and where both currently remain.

In early 2001 another inmate of Florence ADMAX attempted to transmit much of the information developed by Scarpa to the director of the FBI, the President of the United States and the Attorney General. We first became aware of this intelligence when, for purposes entirely unrelated to terrorism, we met with Scarpa at Florence ADMAX in March 2003. In the course of these meetings Scarpa provided to us his contemporaneous (1996-97) handwritten notes of his conversations with Yousef and his terrorist codefendants and copies of FBI Forms 302 (informant reports) recounting the information which he had provided to agents of the FBI.

Scarpa reported conversations with Yousef and with Yousef's codefendants, Wali Khan Amin Shah, Abdul Hakim Murad and Eyad Ismoil. Many of these conversations referenced a person identified by the terrorists as "Bojinga," whom we believe to be Osama Bin Laden.

During his incarceration at the MCC, Scarpa was regularly debriefed by an agent of the FBI who, using the alias "Susan Schwartz," posed as a paralegal employed by Scarpa's attorney, Larry J. Silverman. In addition to these reports, Scarpa, utilizing a mini-camera provided by the warden of the MCC, provided the FBI with photographs of documents ("kites") which Yousef shared with him. Also, with the assistance of Scarpa (acting on behalf of the FBI), Yousef was able to make telephone calls, monitored by the FBI, to active (unincarcerated) members of his terrorist networks via a "patch-through" phone accessed from MCC.

The intelligence provided by Scarpa included specific threats to U.S. airlines, the identification of countries (e.g., England) through which terrorists were entering the United States and testing U.S. security procedures, instructions for smuggling explosive chemicals and detonators (including hiding these in the heels of shoes), and formulas for explosives and for the production of phosgene and mustard gases. Yousef revealed to Scarpa his strong interest in obtaining blank U.S. passports which "his people would only use for one trip to board the planes to be hijacked."

Further intelligence laid out a plan to videotape the killings of hijacking victims, with distribution of these tapes to the media.

One very serious missed opportunity to disrupt a terrorist network here in the United States prior to September 11, 2001, involved Yousef's agreement to meeting in New York between Scarpa's associates (who would have been disguised FBI

E. Letter from Angela Clemente and Dr. Stephen Dresch alerting the 9/11 Commission to the FBI's failure to pursue the leads provided by Scarpa in 1996.

agents) and four active (unincarcerated) terrorists. This meeting never took place because the cognizant FBI agents and assistant U.S. attorneys refused to agree to Yousef's demand that $3,000 be provided to the terrorists. It should be noted that Scarpa provided this assistance to the FBI at considerable risk not only to himself but also to members of his family, whose address he was compelled to provide to Yousef to enhance his credibility.

We have identified the following officials as having been directly involved with Scarpa in his role as an informant, as summarized in the attached "Scarpa-Yousef Intelligence Timeline":

New York Office of the FBI
• Susan Schwartz (alias)
• Pat White
• James Kallstrom

U.S. Attorneys' Offices
• AUSA Michael Garcia (EDNY)
• AUSA Patrick Fitzgerald (EDNY)
• AUSA Valerie Caproni (EDNY)
• U.S. Attorney Mary Jo White (SDNY)

Metropolitan Correctional Center (New York)
• Warden R.M. Reish

Administrative Maximum Security Penitentiary, Florence, Colorado
• Mr. Manly (SIS)

Scarpa's intelligence anticipates "shoe bomber" Richard Reed, the authorization for the 9/11 attacks given by Yousef's uncle, Khalid Shaikh Mohammed, Mohammed Atta's and Al Qaeda's English connections, the single-use passports employed to board the airplanes hijacked on 9/11, and the videotaped death in Pakistan of Wall Street Journal reporter Daniel Pearl. The office of the U.S. Attorney for the Eastern District of New York was initially informed of Scarpa's contacts with Yousef at the MCC and his willingness to serve as an informant by his attorney, Larry J. Silverman (46 Trinity Place, New York 10006; 212-425-1616), who remained informed of Scarpa's activities on behalf of the government.

Although the information provided by Scarpa appears to have been fully documented, especially in Forms 302 prepared by the New York office of the FBI, we do not know if this information was disseminated to other agencies with responsibility for the prevention of terrorist attacks on the United States.

We understand that AUSA Patrick Fitzgerald (EDNY), in a later (sealed) court filing, attested to the credibility, accuracy and value of the terrorism intelligence provided by Scarpa. However, at the time of Scarpa's sentencing AUSA Valerie Caproni denigrated Scarpa's credibility as an informant. Apparently, Scarpa's credibility was perceived to pose a threat in the context of serious questions which had been raised concerning the informant relationship of Scarpa's father, Gregory Scarpa, Sr., with FBI Supervisory Special Agent R. Lindley DeVecchio, the revelation of which might place in jeopardy a number of convictions secured by the office of the U.S. Attorney (EDNY). This latter matter, entirely unrelated to terrorism, remains a subject of continuing, independent investigation.

We strongly advise that the Commission fully examine the terrorism intelligence secured by the government through the informant services of Gregory Scarpa, Jr., and assess the extent to which it was appropriately utilized to reduce the terrorist threat to the United States.

Respectfully submitted,
Angela Clemente
Stephen P. Dresch, Ph.D.
Santrea_143.2@juno.com
706-294-9993

Principal witness: Gregory Scarpa, Jr.
Reg. No. 10099-050
ADX Florence
P.O. Box 8500
5880 State Highway 67 South
Florence, Colorado 81226
719-784-9464
Attachment: Scarpa-Yousef Intelligence Timeline

Appendix III: Documents Relating to the FBI's Knowledge of al Qaeda Prior to 9/11

265A-NY-252802

Continuation of FD-302 of GREGORY SCARPA, JR. , On 3/5/96 , Page 4

getting contact with outside people and YOUSEF believes SCARPA has more contacts on the outside.

According to SCARPA, YOUSEF has indicated that he has four people here. SCARPA described these four people as four terrorists already here in the United States. SCARPA advised YOUSEF has not indicated who these four individuals are or if he is in contact with these four people or how he contacts them. SCARPA does not know if YOUSEF receives or sends any messages from contacts overseas.

According to SCARPA, YOUSEF told SCARPA that if he wanted, SCARPA's family could be sent to an unknown country and people there would take care of his family, treat him like royalty with the red carpet treatment. SCARPA advised YOUSEF has mentioned the Philippines, Jordan and Pakistan, but SCARPA was unsure if any of these were the country he was referring to.

According to SCARPA, YOUSEF implied that another government was involved/assisting YOUSEF. YOUSEF told SCARPA that if he went to this country no one could touch him.

SCARPA advised YOUSEF received newspapers and may get letters, and occasionally talks on the phone.

SCARPA indicated he felt YOUSEF had a hard time getting to talk to his friends, and therefore, wants to get SCARPA's people to talk to his people and get assistance.

According to SCARPA, YOUSEF told him that if he felt SCARPA was serious, YOUSEF would ask him to pay $3,000. YOUSEF did not say what the $3,000 would be used for. SCARPA advised that YOUSEF brought up the issue of money several times. SCARPA said it is important to YOUSEF to accumulate money. SCARPA advised the $3,000 would show YOUSEF that SCARPA was serious and ready to deal.

When SCARPA was asked if he had any knowledge of YOUSEF committing acts in furtherance of a kidnapping in the United States, SCARPA advised YOUSEF told him it had worked in another country, but SCARPA could provide no specifics about anything currently being planned in the United States.

SCARPA advised that YOUSEF is teaching him to make a bomb by passing the information on "kites". SCARPA remembered the words nitroglycerine, acetone and 16-24 hours ice bath.

A. An FBI 302 detailing the existence of Yousef's terror cell ("four terrorists already here in the United States"): March 5, 1996.

265A-NY-258172

Continuation of FD-302 of GREGORY SCARPA, JR. (PROTECT IDENTITY) On 09/09/96 Page

The fourth note pertained to the events taking place on Sunday, September 8th. YOUSEF told SCARPA (Protect Identity) that BOJINGA contacted him during his trial and said that his people are going to prepare to hijack an airplane and kidnap a United States Ambassador at the same time if he is convicted. BOJINGA told YOUSEF that the plan would be carried out within one year of YOUSEF's conviction. YOUSEF advised that BOJINGA said that after they hijack the airplane and kidnap the Ambassador that BOJINGA will ask for the release of YOUSEF and all of the defendants of the World Trade Center and the defendants in the SHIEK's case. SCARPA (Protect Identity) advised that YOUSEF mentioned eleven people total. BOJINGA told YOUSEF that after they hijack the airplane they are going to contact the U.S. Embassy and ask to be given a phone number to speak with YOUSEF. YOUSEF said in exchange for speaking with YOUSEF BOJINGA will release all children and mothers off of the plane. YOUSEF stated that BOJINGA will ask YOUSEF for the names of the people he wants released besides his codefendants. YOUSEF stated that the hijacking could include more than one airplane. YOUSEF said that the airplanes would be hijacked and forced to land in either Afghanistan, Sudan, or Yemen. BOJINGA asked YOUSEF if YOUSEF can arrange through his contacts here in the United States for two to four passports. YOUSEF said that he has to contact BOJINGA to let him know if he can help. YOUSEF asked SCARPA (Protect Identity) if SCARPA (Protect Identity) can get him the passports. If so, BOJINGA will get an address in Iran or Libya where he can send the passports. YOUSEF would like SCARPA (Protect Identity) to provide an address so that BOJINGA can send photographs for the passports.

B. An FBI 302 detailing plans to hijack one or more airplanes: September 8, 1996.

FEDERAL BUREAU OF INVESTIGATION

Date of transcription 12/26/96

Confidential Source (CS) GREGORY SCARPA, JR. (Protect Identity) was interviewed at the Metropolitan Correctional Center (MCC) at 150 Park Row, New York, New York (NY). He provided the following information:

CS transferred three (3) pages of handwritten notes to the interviewing Agent. CS advised that the notes were self-written. The first two pages pertained to the events taking place on December 18, 1996. CS asked YOUSEF to reveal the address that the passports were to be sent. YOUSEF responded that he will not give out the address until he finds out whether CS associates will send them or not. CS questioned YOUSEF whether the address is located in the United States. YOUSEF responded that the address is an Iranian address. CS asked whether YOUSEF was having the passports sent to his parents address. YOUSEF responded "No", and that a temporary address was set up in Iran for this purpose.

CS questioned YOUSEF how he sends and receives his messages. YOUSEF responded that he will reveal his method in a few weeks. YOUSEF told CS that "You'll be real surprised" and "You'll be shocked". YOUSEF said for his own reasons he cannot tell his method yet.

CS advised that he noticed a few sores on MURAD's face during conversation. CS questioned MURAD regarding the sores. MURAD responded that they were from the high fever he had while he was sick.

CS advised that, prior to meeting with the interviewing Agent on December 18, 1996, he had a conversation with YOUSEF regarding co-operating witnesses on the L-unit. YOUSEF was curious as to what kind of deals the Government makes with the witnesses. CS explained that it depends on the information obtained and how well the witness does at trial. CS told YOUSEF that he should make a deal with the Government. CS advised that YOUSEF should give up "BOJINGA" or BIN LADEN. YOUSEF responded that the Government would never go after BIN LADEN because the Government knows that within one week of capturing BIN LADEN twelve U.S. airplanes would be blown up.

Investigation on 12/26/96 at New York, New York

File # 265A-NY-258172 Date dictated 12/26/96

by SA [redacted]

(19)

C. An FBI 302 detailing Yousef's threat that "twelve U.S. airplanes would be blown up" if the U.S. should capture bin Laden: December 26, 1996.

Declassified and Approved
for Release, 10 April 2004

Bin Ladin Determined To Strike in US

Clandestine, foreign government, and media reports indicate Bin Ladin since 1997 has wanted to conduct terrorist attacks in the US. Bin Ladin implied in US television interviews in 1997 and 1998 that his followers would follow the example of World Trade Center bomber Ramzi Yousef and "bring the fighting to America."

After US missile strikes on his base in Afghanistan in 1998, Bin Ladin told followers he wanted to retaliate in Washington, according to a service.

An Egyptian Islamic Jihad (EIJ) operative told an service at the same time that Bin Ladin was planning to exploit the operative's access to the US to mount a terrorist strike.

The millennium plotting in Canada in 1999 may have been part of Bin Ladin's first serious attempt to implement a terrorist strike in the US. Convicted plotter Ahmed Ressam has told the FBI that he conceived the idea to attack Los Angeles International Airport himself, but that Bin Ladin lieutenant Abu Zubaydah encouraged him and helped facilitate the operation. Ressam also said that in 1998 Abu Zubaydah was planning his own US attack.

Ressam says Bin Ladin was aware of the Los Angeles operation.

Although Bin Ladin has not succeeded, his attacks against the US Embassies in Kenya and Tanzania in 1998 demonstrate that he prepares operations years in advance and is not deterred by setbacks. Bin Ladin associates surveilled our Embassies in Nairobi and Dar es Salaam as early as 1993, and some members of the Nairobi cell planning the bombings were arrested and deported in 1997.

Al-Qa'ida members—including some who are US citizens—have resided in or traveled to the US for years, and the group apparently maintains a support structure that could aid attacks. Two al-Qa'ida members found guilty in the conspiracy to bomb our Embassies in East Africa were US citizens, and a senior EIJ member lived in California in the mid-1990s.

A clandestine source said in 1998 that a Bin Ladin cell in New York was recruiting Muslim-American youth for attacks.

We have not been able to corroborate some of the more sensational threat reporting, such as that from a service in 1998 saying that Bin Ladin wanted to hijack a US aircraft to gain the release of "Blind Shaykh" 'Umar 'Abd al-Rahman and other US-held extremists.

— Nevertheless, FBI information since that time indicates patterns of suspicious activity in this country consistent with preparations for hijackings or other types of attacks, including recent surveillance of federal buildings in New York.

The FBI is conducting approximately 70 full field investigations throughout the US that it considers Bin Ladin–related. CIA and the FBI are investigating a call to our Embassy in the UAE in May saying that a group of Bin Ladin supporters was in the US planning attacks with explosives.

For the President Only
6 August 2001

D. The notorious PDB memo: August 6, 2001.

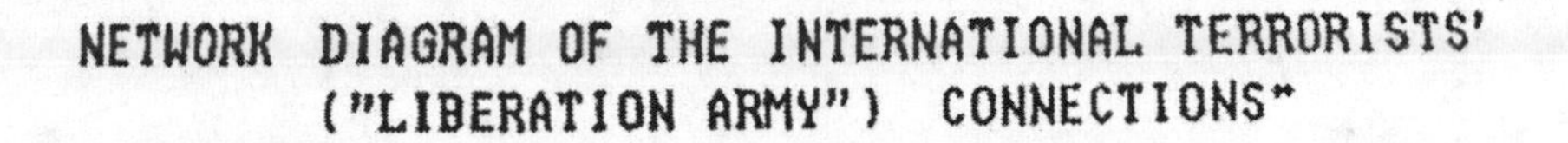

NETWORK DIAGRAM OF THE INTERNATIONAL TERRORISTS'
("LIBERATION ARMY") CONNECTIONS*

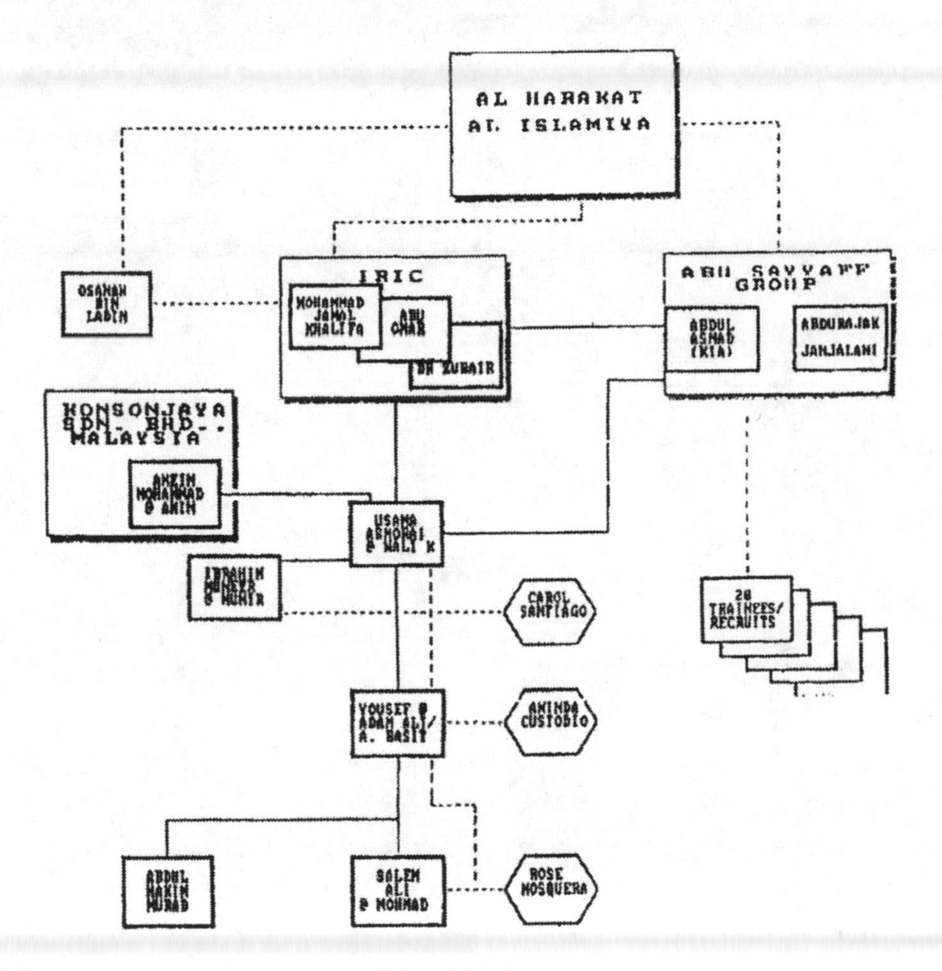

* BASED ON DATA OBTAINED FROM THE OPERATION

E. The flow chart Col. Rodolfo Mendoza sent the FBI in 1995, detailing the connection between bin Laden and the deadly Abu Sayyaf terror group.

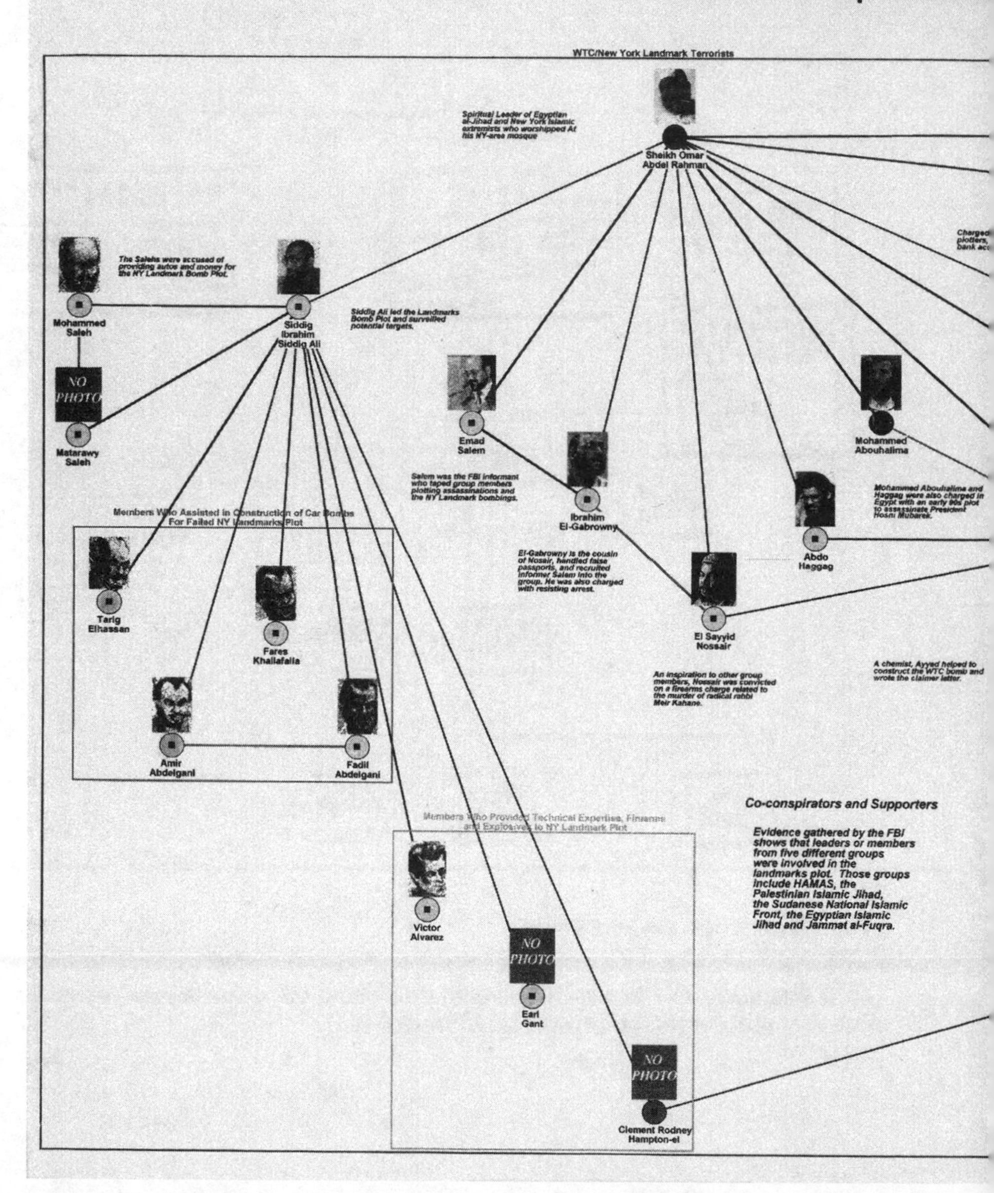

F. The DIAC link chart showing connections from the Yousef-Rahman New York cell to bin Laden, Walih El-Hage, and Ali Mohammed in Africa, and al Qaeda (right): August 10, 1998.

ew York Landmarks Bomb Plot:
and International Linkages

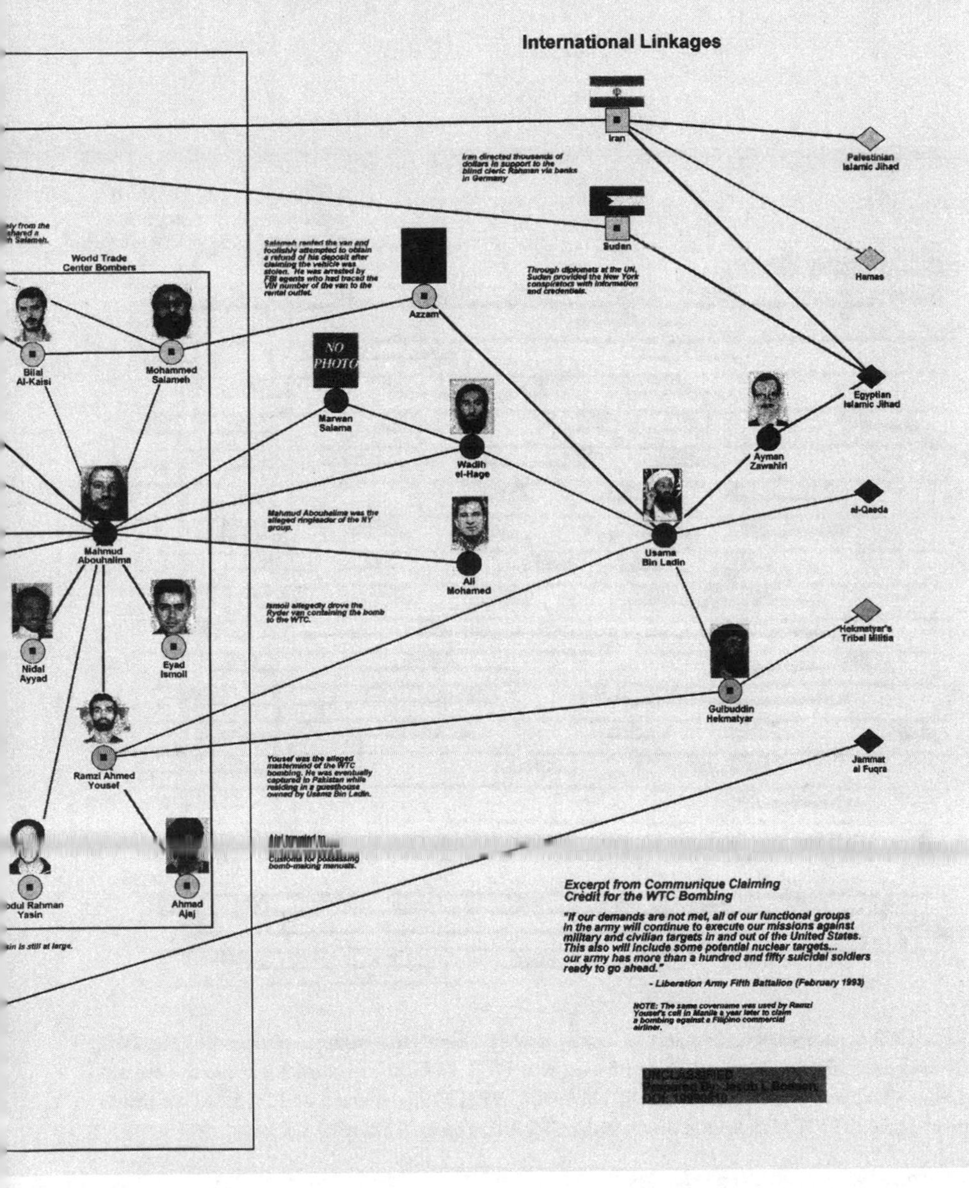
International Linkages
Iran
Iran directed thousands of dollars in support to the blind cleric Rahman via banks in Germany
Palestinian Islamic Jihad
Sudan
Through diplomats at the UN, Sudan provided the New York conspiratocs with information and credentials.
Hamas
Egyptian Islamic Jihad
World Trade Center Bombers
Salameh rented the van and foolishly attempted to obtain a refund of his deposit after claiming the vehicle was stolen. He was arrested by FBI agents who had traced the VIN number of the van to the rental outlet.
Azzam
Bilal Al-Kaisi
Mohammed Salameh
NO PHOTO
Marwan Salama
Wadih el-Hage
Ayman Zawahiri
Mahmud Abouhalima was the alleged ringleader of the NY group.
Mahmud Abouhalima
Ali Mohamed
Usama Bin Ladin
al-Qaeda
Ismoil allegedly drove the Ryder van containing the bomb to the WTC.
Nidal Ayyad
Eyad Ismoil
Hekmatyar's Tribal Militia
Gulbuddin Hekmatyar
Ramzi Ahmed Yousef
Yousef was the alleged mastermind of the WTC bombing. He was eventually captured in Pakistan while residing in a guesthouse owned by Usama Bin Ladin.
Jammat al Fuqra
Customs for possessing bomb-making manuals.
dul Rahman Yasin
Ahmad Ajaj
sin is still at large.
Excerpt from Communique Claiming Crèdit for the WTC Bombing
"If our demands are not met, all of our functional groups in the army will continue to execute our missions against military and civilian targets in and out of the United States. This also will include some potential nuclear targets... our army has more than a hundred and fifty suicidal soldiers ready to go ahead."
- Liberation Army Fifth Battalion (February 1993)
NOTE: The same covername was used by Ramzi Yousef's cell in Manila a year later to claim a bombing against a Filipino commercial airliner.

Appendix IV: Documents Relating to the Crash of TWA 800

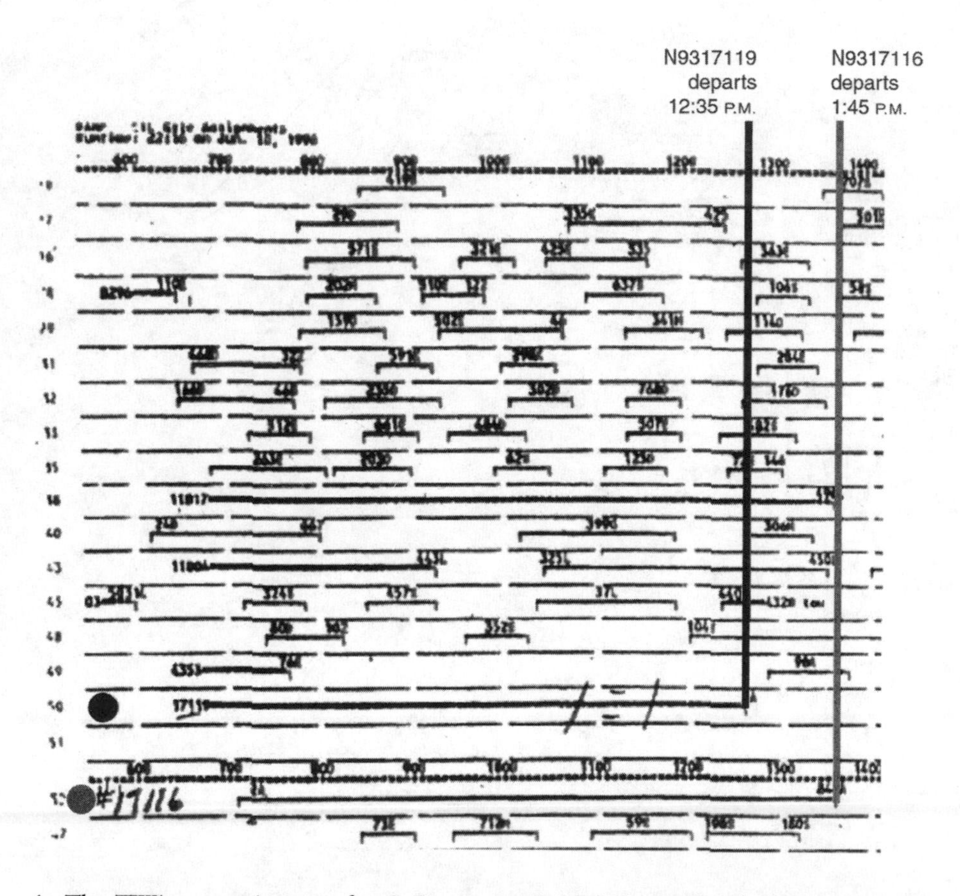

A. The TWA gate assignment for C Concourse, St Louis International Airport, June 10, 1996—the day Officer Herman Burnett used the TWA 747–100 to train his explosive-detecting dogs. The plane that later became TWA 800, 9317119, departed at 12:35 P.M.; a nearly identical plane, 9317116, lingered across the concourse at gate 52 until 1:45 P.M.

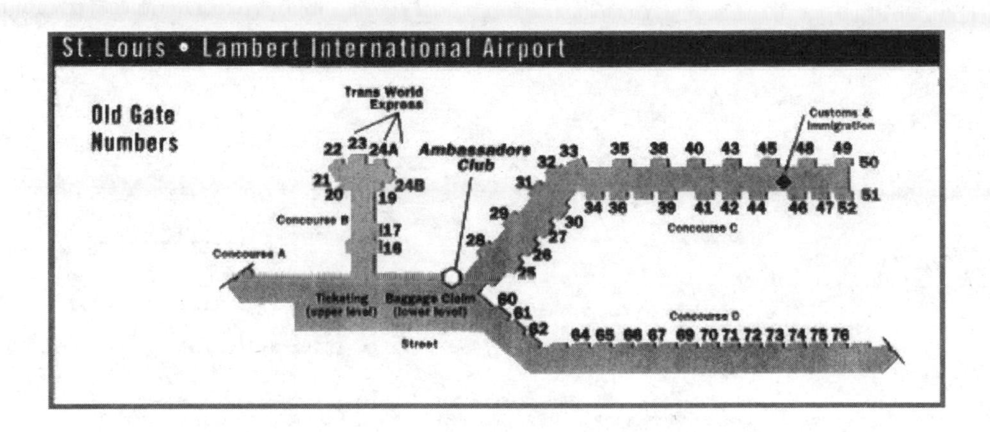

B. A map of Concourse C as it was in 1996, showing gates 50 and 52 across the concourse from each other (above). Below: Researcher Tom Shoemaker's diagram showing the placement of Officer Burnett's canine exercise explosives (indicated by arrows and dots) and the residue found on the wreckage of TWA 800 (the wide oval near door 2).

REPORTED PLACEMENT POINTS OF CANINE EXERCISE EXPLOSIVES
REPORTED EXPLOSIVE RESIDUES ON RECOVERED WRECKAGE

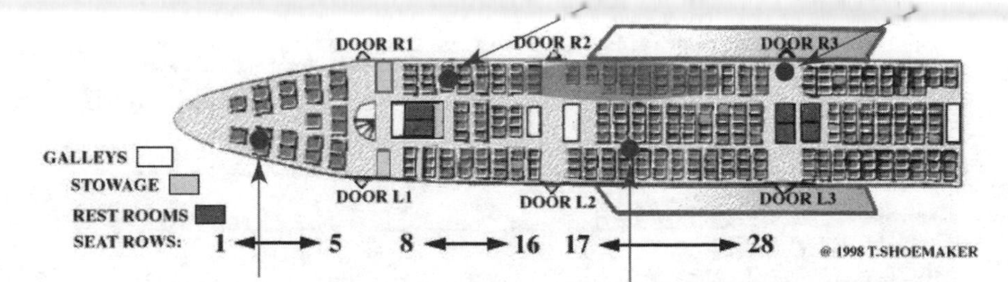

Appendix V: Documents Relating to the DeVecchio OPR

FEDERAL BUREAU OF INVESTIGATION

Date of transcription 2/7/94

LAWRENCE MAZZA was interviewed by properly identified Special Agents of the Federal Bureau of Investigation. MAZZA then provided the following information, as well as information not included here:

SCARPA SR. told MAZZA that he received information from a source in law enforcement, and a source from the ORENA faction. One of the sources that supplied information on a regular basis was referred to as "THE GIRLFRIEND". The source would call him at home, as well as beep him, and was always referred to by SCARPA SR. as the "GIRLFRIEND". SCARPA SR.'s wife, LINDA, also referred to the source as "THE GIRLFRIEND" when giving messages to SCARPA SR. regarding phone calls. On occasions when the source beeped SCARPA SR., he would stop whatever he was doing in order to return the call. When in his own home, SCARPA SR. always returned the calls to the "GIRLFRIEND" using the telephone in the basement, which was registered to MARYANN TURSI. SCARPA Sr. also received messages regularly to call "THE GIRLFRIEND" at the "store", which referred to the source's place of business.

The information that SCARPA SR. received through his source(s) included, but was not limited to, the address of VICTOR ORENA's girlfriend's home in Queens, New York, including descriptive information that it was a white, two family home with aluminum siding. Shortly after SCARPA SR. is shot at, he learns that the panel truck used in the murder attempt was rented from Queens, New York. JOSEPH RUSSO's, an associate of WILLIAM CUOTOLO, address on 74th Street near 12th or 13th Avenue, as well as SALVATORE MICIATTA and JOSEPH SCOPO's addresses. He also received information regarding the scheduled drug arrest of GREGORY SCARPA JR. and his crew, and the scheduled Credit Card arrest of SCARPA SR.. He received a copy of the Complaint issued against himself, and was told that his problems with the Law might disappear if he stayed out of trouble.

On one occasion, MAZZA was in SCARPA SR.'s house when he received a telephone call from "THE GIRLFRIEND" on the basement telephone. After the call, he told MAZZA that the "GIRLFRIEND" said that the members of the ORENA faction were very close to killing him.

Investigation on 1/7/94-2/7/94 at Undisclosed File # 281A-NY-214955

by Maryann C. Walker. Christopher M. Favo Date dictated 2/7/94

This document contains neither ...

A. The infamous missing "girlfriend" FBI 302.

(Ю 31/31)

FEDERAL BUREAU OF INVESTIGATION

Precedence: IMMEDIATE **Date:** 04/10/1996

To: DIRECTOR, FBI **Attn:** ASSISTANT DIRECTOR, INSPECTION DIVISION

From: ADIC, NEW YORK
Contact: JAMES J. ROTH

Approved By: ROTH JAMES J

Drafted By: ROTH JAMES J:jjr

File Number(s): 263- (Pending)

Title: SUPERVISORY SPECIAL AGENT R. LINDLEY DEL VECCHIO OPR MATTER

NY requests that whatever investigation is to be conducted as a result of this letter be conducted expeditiously, with the results provided to DOJ, and that DOJ be strenuously pressed to provide a prosecutive opinion regarding this matter so that it may be resolved. NY believes, based on the investigative results to date and assuming this latest information does not change the result, that there is insufficient evidence to take prosecutive action against SSA DelVecchio. The failure of the DOJ to provide a prosecutive opinion or for the FBI to administratively resolve this matter continues to have a serious negative impact on the government's prosecutions of various LCN figures in the EDNY and casts a cloud over the NYO.

B. Memo from ADIC James Kallstrom and PLO James Roth to FBI Director Louis Freeh urging the resolution of the OPR against DeVecchio, which "casts a cloud over the NYO" (above). Below: The DAR that proves that Det. Joe Simone was called back to 26 Federal Plaza just before Gregory Scarpa Sr.'s murder of Nicholas Grancio.

INVESTIGATORS DAILY ACTIVITY REPORT PD439-156 (REV. 2-77) Print Legibly or Type

OVERTIME Straight Time	HOURS 5	MINUTES 45	SHIELD NUMBER 1945	TAX REG. NUMBER 873611	RANK, NAME DET JOE SIMONE			
Last Report Submitted/Dates Reasons Report Not Submitted (R.D.O., Vacation, etc.) MON 1/6				Tour TUES	Day 1245X2100	Date of Activity 1/7/92	Command JCB/OCID#4	

Time From	Time To	Activity: Include members assigned with, arrest numbers, property seized, etc.	Area or Major Case Number	Expense Items	Amount Expended
1245	1330	S/D Plant # 26 - Columbo Task Force			
1330	1400	ERT 26 Fed Pl. w/ Maggiore 1/116-GJM			
1400	1600	A/A w/o Re: Confer w/ C. Favo.			
1600	1630	ERT V/o Village Rd N. + McD. Ave Bklyn w/a 1/a			
1630	1730	A/A w/a Re: Grancio Homicide			
1740	1750	ERT ~~[illegible]~~ CI Hosp. w/a 1/a			
1750	1820	A/A w/a Re: Interview Anthony Bianco			
1820	1825	ERT 61 Pct w/a 1/a			
1825	2000	A/A w/a Re: Interview Joey Tolino			

Appendix VI: Membership of the 9/11 Commission Family Steering Committee

Carol Ashley
mother of Janice Ashley, 25

Kristen Breitweiser
wife of Ronald Breitweiser, 39

Patty Casazza
wife of John F. Casazza, 38

Beverly Eckert
wife of Sean Rooney

Mary Fetchet
mother of Bradley James Fetchet, 24

Monica Gabrielle
wife of Richard Gabrielle

Bill Harvey
husband of Sara Manley Harvey, 31

Mindy Kleinberg
wife of Alan Kleinberg, 39

Carie Lemack
daughter of Judy Larocque

Sally Regenhard
mother of Christian Michael Otto Regenhard, 28

Lorie Van Auken
wife of Kenneth Van Auken, 47

Robin Wiener
sister of Jeffrey Wiener, 33

Family Steering Committee
265 Sunrise Highway
Suite 1 #334
Rockville Centre, NY 11570
www.911independentcommission.org

NOTES

INTRODUCTION

1 Peter Lance, *1000 Years for Revenge: International Terrorism and the FBI—The Untold Story* (New York: ReganBooks, 2003).

2 The former Democratic Senator from Georgia was one of the original 9/11 Commissioners. In the first year of the Commission's investigation Cleland became the harshest critic on the panel of attempts by the Bush Administration to limit the disclosure of documents to the Commission. In December, 2003, he left the Commission after he was appointed by President Bush to fill a Democratic slot on the Export-Import bank board. Philip Shenon, "Ex-Senator Will Soon Quit 9/11 Panel, Leaving Gap for Victims' Advocates," *New York Times,* December 5, 2003; Eric Boehlet, "The President Ought to Be Ashamed: Former Sen. Max Cleland Blast Bush's 'Nixonian' Stonewalling of the 9/11 Commission," *Salon.com,* November 21, 2003.

3 Excerpt of C-Span hearing coverage, http://peterlance.com/commissionhearing.htm.

4 Public Law 107-306, Title VI: National Commission on Terrorist Attacks upon the United States, Sec. 602 (3)116 Federal Statutes, p. 2408, November 27, 2002.

5 Richard C. Shelby, "September 11th and the Imperative of Reform in the U.S. Intelligence Community: Additional Views of Senator Richard C. Shelby, Vice Chairman, Senate Select Committee on Intelligence," December 10, 2002.

6 Gail Sheehy, *Middletown, America: One Town's Passage from Trauma to Hope* (New York: Random House, 2003).

7 The official death toll as of July, 2004, was 2,973. 9/11 Commission Report (New York: Norton, 2004), p. 311.

8 Author's interview with Lori van Auken, May 24, 2004.

9 David E. Rosenbaum, "Threats and Responses: The Commissioners—For Members of Panel, Past Work Becomes an Issue in the Present Hearings," *New York Times,* April 14, 2004.

10 Lance, *1000 Years for Revenge,* pp. 4–5.

11 Public Law 107-306, Title VI.

12 Sara Klugler, "Protesters Overshadow Giuliani's Testimony," *Toronto Star,* May 19, 2004, www.thestar.com.

13 Author's interview with Angela Clemente, researcher, Forensic Intelligence, April 11, 2004.

PART I

1. THE FBI'S KILLING MACHINE

1 *U.S. v. Cecil Ray Price et al.,* Criminal Action Number 5291, U.S. Court for the Southern District of Mississippi, October 7, 1968.
2 Ibid., trial transcript p. 765.
3 Ibid., p. 957.
4 Andrew Jacobs, "The Southern Town Struggles with a Violent Legacy," *New York Times,* May 29, 2004.
5 Author's interview with Judge W. O. Chet Dillard, June 11, 2004.
6 FBI AIRTEL re: Top Echelon Informant, June 6, 1962; Payment authorization from Asstant Director Evens to Director J. Edgar Hoover.
7 Greg B. Smith and Jerry Capeci, "Mob, Mole & Murder," *New York Daily News,* October 30, 1994.
8 Anthony Villano with Gerald Astor, *Brick Agent: Inside the Mafia for the FBI* (New York: Quadrangle, 1977).
9 Testimony of Marty Light, Gregory Scarpa Sr.'s former attorney, at hearings before the President's Commission on Organized Crime, Washington, D.C., January 29, 1986.
10 On November 21, 2003, Alphonse Persico was sentenced to thirteen years in prison for racketeering, extortion, and loan sharking. Judge Reena Raggi, presiding at the U.S. Court for the Eastern District of New York, described Persico, forty-nine, as "a very dangerous man." John Marzulli, "Colombo Big Gets Thirteen Years," *New York Daily News,* November, 22, 2003.
11 *U.S. v. Orena,* 92 CR 351, affirmation of Flora Edwards.
12 FBI Memorandum from Asst. Director re: Scarpa's recruitment for the Mississippi interrogation of the Klansman, January 21, 1966.
13 Sandra Harmon, "Scarpa's Mistress," book proposal, May 1, 2000.
14 W. O. Chet Dillard, *Clear Burning: Civil Rights, Civil Wrongs* (Jackson, MS: Persimmon Press, 1992).
15 Author's interview with Judge W. O. Chet Dillard, June 11, 2004.
16 Frederic Dannen, "The G-Man and the Hit Man," *New Yorker,* December 16, 1996.
17 Unique Way of Solving Mystery, *Pittsburgh Post Gazzette,* December 1, 1998; Tom Robbins and Jerry Capeci, "FBI Used Wiseguy to Crack KKK Man," *New York Daily News,* June 21, 1994.
18 Jerry Mitchell, "Feds to Aid in '64 Slayings Case," *Clarion-Ledger,* May 29, 2004.
19 Andrew Jacobs, "The Southern Town Struggles with a Violent Legacy," *New York Times,* May 29, 2004.
20 Dannen, "The G-Man and the Hit Man."
21 Author's interview with Alan Futerfas, May 24, 2004.
22 Sworn Affidavit of Gregory Scarpa Jr., July 30, 2002.
23 Affirmation of Flora Edwards; Author's interview with Flora Edwards, April 26, 2004; Author's interview with Det. Joseph Simone, NYPD (ret.), April 30, 2004.
24 Affidavit of Gregory Scarpa Jr., April 29, 1999; Affidavit of Gregory Scarpa Jr., 2002.
25 FBI 302 reports re: interview of Christopher M. Favo, November 16, 1995 and December 8, 1995; Alan J. Futerfas and Ellen Resnick, "Final Submission in Support of Victor J. Orena and Pasquale Amato's Motion for a New Trial," October 16, 1996. *U.S. v. Anthony Russo et al.;* FBI 302 interrogation of Larry Mazza, April 28, 1994.

26 Bill Moushey, "Switching Sides: Federal Agents Sometimes Fall Prey to the Lurid Lifestyles of Their Informants," *St. Louis Post-Dispatch,* December 1, 1998.

27 Affirmation of Flora Edwards; Exhibit M. *Victor J. Orena and Pasquale Amato v. U.S.,* Testimony of R. Lindley DeVecchio; Cross Examination by Gerald Shargel, Ibid., transcript pp. 158–166.

28 Pasquale Amato, *Victory Orena v. CV-96-1461; CV-96-1474,* Affirmation of Flora Edwards, Exhibit M., p. 9 January 7, 2004.

29 Author's interview with Alan Futerfas, May 24, 2004; Author's interview with Flora Edwards, April 27, 2004; Dannen, "The G-Man and the Hit Man."

2. THE MOZART OF TERROR

1 KSM was captured in Rawalpindi, Pakistan, on March 3, 2003; "Alleged 9-11 Mastermind Nabbed," CBSNews.com, March 3, 2003.

2 "America's Most Wanted," *Newsweek,* July 4, 1994.

3 FBI 302 interrogation of Ramzi Yousef, February 13, 1995.

4 *U.S. v. Ramzi Ahmed Yousef,* S10 93 Cr. 180 (KTD), indictment and "want" poster Department of Justice, U.S. State Department Office of Diplomatic Security; Mary Ann Weaver, "Children of the Jihad," *New Yorker,* June 12, 1995. While in New York during the fall of 1992 as he built the urea-nitrate World Trade Center device, Yousef went by the name "Rashed."

5 Greg Scarpa Jr., "kites" from notes passed by Ramzi Yousef on Nine South of the Metropolitan Correctional Center, May–December 1996; FBI 302s authenticating Scarpa Jr.'s intelligence reports, May–December 1996.

6 David B. Ottoway and Steve Coll, "Retracing the Steps of a Terror Suspect: Accused Bomb Builder Tied to Many Plots," *Washington Post,* June 5, 1995.

7 FBI 302 interrogation of Ramzi Yousef, February 7–8, 1995, pp. 12–13.

8 In March, 1994, Yousef was suspected in the aborted bombing of the Israeli Embassy in Bangkok, where a 2000-pound ammonium nitrate diesel fuel bomb was found abandoned in a truck along with the body of the driver: "Massive Bomb, Dead Body Found in Truck," *Bangkok Post,* March 18, 1994. In June, 1994 Yousef was tied to the bombing of the Imam Reza Shrine in the city of Mashad, Iran's most sacred Shiite place of worship. Twenty six died and more than 200 were injured: John Miller, Michael Stone, and Chris Mitchell, *The Cell: Inside the 9/11 Plot and Why the FBI and CIA Failed to Stop It* (New York: Hyperion, 2002), p. 120. The pipe bombs discovered after a fire in his bomb factory at the Doña Josefa Apartments in Manila on January 6, 1995, revealed Yousef's plot to plant the devices along the route of Pope John Paul II's motorcade January 12, 1995: FBI Laboratory inventory: Manila Air; Bombing of PAL FLIGHT 434 December 12, 1994, and Subsequent Investigation 265A-IIN-12924; obtained by the author.

9 *U.S. v. Ramzi Ahmed Yousef et al.,* May 29, 1996.

10 Author's interview with PNP Col. Rodolfo B. Mendoza, April 19, 2002.

11 "Cold Blooded," PNP Superintendent Samuel Pagdilao, Philippines National Police, Nicholas Cumming-Bruce, "Manila Turns Terrorist Over to U.S." *Guardian,* April 15, 1995; "Diabolical," Simon Reeve, "The New Jackals," (Boston: Northeastern University Press, 1999) FBI Asst. Director Neal Herman, p. 249; "Evil Genius," Senior Pakistani Intelligence Officer, Ibid., Reeve.

12 *U.S. v. Ramzi Yousef and Eyad Ismoil.*
13 *U.S. v. Ramzi Yousef et al.*, May 29, 1996.
14 M. A. Farber, "Kahane Trial Sets Off Squabbles by Lawyers," *New York Times*, December 9, 1991.
15 *U.S. v. Mohammed A. Salameh et al.*, S593CR.180 (KTD), Kevin T. Duffy.
16 *U.S. v. Omar Abdel Rahman et al.*, October 9, 1995.
17 Steven Emerson, *American Jihad* (New York: Free Press, 2003); Yohah Alexander and Michael Swetnam, *Osama bin Laden's al Qaida: Profile of a Terrorist Network*, p. 39.
18 Steven Emerson, *American Jihad* (New York: Free Press, 2003); Yohah Alexander and Michael Swetnam, *Osama bin Laden's al Qaida: Profile of a Terrorist Network*, p. 39.
19 *U.S. v. Osama bin Laden*, testimony of Jamal Ahmed al Fadl, February 6, 2001.
20 Dr. Ayman al-Zawahiri and the late Mohammed Atef. Atef, al Qaeda's former military leader, was linked by U.S. investigators to the October 1993, attack on U.S. forces in Somalia, which led to the downing of two Blackhawk helicopters. He was reportedly killed in the early days of the U.S. invasion of Afghanistan in November 2001.
21 Stephen Emerson, "Abdullah Azzam: The Man before Osama bin Laden," *Journal of Counterterrorism and Security International*, vol. 5, no. 3, Fall 1998.
22 Stephen Engelberg, "One Man and a Global Web of Violence," *New York Times*, January 14, 2001.
23 Questions about the Sheikh's U.S. entry surfaced publicly in early 1993 and continued throughout the year following the bombing of the World Trade Center on February 26, 1993, and the Sheikh's surrender to federal agents on July 3, 1993. Chris Hedges, "A Cry of Islamic Fury, Taped in Brooklyn for Cairo," *New York Times*, January 7, 1993; Timothy Carney and Mansoor Ijaz, "Intelligence Failure? Let's Go Back to Sudan," *Washington Post*, June 30, 2002; James C. McKinley Jr., "Islamic Leader on U.S. Terrorist List in Brooklyn," *New York Times*, December 16, 1990.
24 Ronald Sullivan, "In Eye of Storm, Jury Selection Begins in Kahane Killing," *New York Times*, November 5, 1991.
25 Ralph Blumenthal, "Clues Hinting at Terror Ring Were Ignored," *New York Times*, August 27, 1993.
26 Author's interview with U.S. Postal Inspector Frank Gonzalez (ret.), January 3, 2003.
27 Author's interview with Det. Lou Napoli NYPD (ret.), January 13, 2003.
28 Engelberg, "One Man and a Global Web of Violence."
29 Alison Mitchell, "After Blast, New Interest in Holy-War Recruits in Brooklyn," *New York Times*, April 11, 1993.
30 Susan Schmidt, "1998 Memo Cited Suspected Hijack Plot by bin Laden," *Washington Post*, July 18, 2004.
31 Author's interview with Special Agent Len Predtechenskis FBI (ret.), August 27, 2002.
32 Lance, *1000 Years for Revenge*, pp. 55–57.
33 Author's interview with confidential FBI source.
34 *U.S. v. Omar Abdel Rahman et al.*, March 7, 1995.
35 *U.S. v. Omar Abdel Rahman et al.*, March 8, 1995.
36 Author's interview with Len Predtechenskis, August 27, 2002; Author's interview with FBI Principal Legal Officer Jim Roth, May 30, 2003.
37 Author's interview with Carson Dunbar June 6, 2003.
38 Author's interview with Len Predtechenskis August 27, 2002.
39 9/11 Commission Staff Statement no. 15, June 16, 2004.

40 *U.S. v. Ramzi Ahmed Yousef and Eyad Ismoil,* September 2, 1997.

41 Author's interview with Len Predtechenskis, August 27, 2002.

42 Author's interview with FBI Agent Julian Stackhouse (ret.), June 3, 2003.

43 Richard Bernstein, "Testimony in Bomb Case Links Loose Ends," *New York Times,* January 19, 1994

44 Engelberg, "One Man and a Global Web of Violence."

45 Author's interview with U.S. Postal Inspector Frank Gonzalez (ret.), January 3, 2003.

46 Author's interview with Eve Bucca, June 17, 2002, wife of FDNY fire marshal Ronnie Bucca, an Army Reservist with Top Secret security clearance who worked at the Defense Intelligence Analysis Center (DIAC) at Bolling Air Force Base in Washington. Bucca, who obtained a copy of Yousef's credit letter addendum, told dozens of people in the FDNY and New York law enforcement community that he was certain that Ramzi Yousef would one day return to New York to attack the Twin Towers again. On the morning of September 11, 2003, as recounted in *1000 Years for Revenge,* Bucca and Battalion Chief Oreo Palmer made it up to the seventy-eighth floor of the South Tower near the area where United Flight 175 had crashed. Radio broadcasts from that day suggest that they charged a standpipe hose and were fighting the flames when the Tower collapsed beneath them. Bucca remains the first and only Fire Marshal in the history of the FDNY killed in the line of duty. *Note:* Bucca's early sense of Yousef's prediction was vindicated two days after 9/11 in a *Washington Post* story by Peter Sleven and Walter Pincus, "Attacks Studied Mistakes in Previous Assaults," *Washington Post,* September 13, 2001.

3. THE SUICIDE-HIJACK PLOT

1 "chocolate" was an arbitrary code word that Yousef chose to represent explosives; FBI 302 interrogation of Abdul Hakim Murad, April 12–13, 1995 p. 16.

2 *U.S. v. Ramzi Yousef et al.,* August 26, 1996, p. 5110.

3 Christopher, Wren "Jury Convicts 3 in Conspiracy to Bomb Airliners." *New York Times.* September 5, 1996.

4 Alpha Tango Flying School in Bern Stages, Texas, near San Antonio, Richmore Flying School in Schenectady, New York, Coastal Aviation in New Bern, North Carolina, and the California Aeronautical Institute in Red Bluff California.

5 The difference in spelling between the most commonly used form of bin Laden's name "Osama" and the form used by the FBI and the Justice Department is worthy of note. The variations are a function of transliteration. Normal Goldstein, the stylebook editor for the AP was quoted as saying "Arab vowels commonly become A, I or U; E and O don't really exist except for personal preference. And since we're not going to ask Osama bin Laden his preference, Osama—more often than not—is spelled with an 'O.'" Robert K. Elder, "Usama, Osama? Tracking Suspects Spells Confusion," *Chicago Tribune,* October 22, 2001. *Note:* While the discussion seems somewhat arcane, the variation in Islamic names makes tracking members of al Qaeda (aka al-Qaida) extremely difficult, especially considering the number of aliases used by terrorists. Even journalists attempting to untangle the Gordian knot differ as to the proper spelling and pronunciation. In this book we have referred to "The Red" as Mahmud Abouhalima, the same spelling used in his federal indictment for the WTC

bombing. But in *The War Against America* scholar Laurie Mylroie spells his last name Abu Halima (loosely translated "father of Halima").

6 In January of 1995, shortly after his capture by Philippines National Police in Manila, Shah escaped from a bungalow where he was being housed at Camp Crame, a PNP base. Then on February 7, 1996, after being captured and returned to the U.S. for trial, he broke through a cage covering the rooftop recreation area at the Metropolitan Correctional Center in New York and made his way onto a ledge twelve stories about the city before being captured; *U.S. v. Ramzi Ahmed Yousef et al.*, August 7, 1996.

7 PNP Person Report on Mohammed Jamal Khalifa, August 25, 1996; PNP Intelligence Command "Osama bin Laden," April 8, 1994; PNP Intelligence Command: "Summary of Activities Leading up to Pope's Visit, Dona Josefa Fire and Aftermath," February 27, 1995; PNP Intelligence Command: "Tactical Interrogation Report: Abdul Hakim Murad" February 8, 1995; Rodolfo J. Garcia, Police Superintendent: "Recommendations for Meritorious Promotion," June 8, 1995; FBI 302 interrogation of Ramzi Yousef aka Abdul Basit Mahmud Abdul Karim, February 7–8, 1995; Ottoway and Coll, "Retracing the Steps of a Terror Suspect: Accused Bomb Builder Tied to Many Plots"; Mark Fineman and Richard C. Paddock, "Indonesian Cleric Tied to '95 Anti U.S. Plot," *Los Angeles Times,* February 7, 2002.

8 Founded by Abdurajak Janjalani, another veteran of the Afghan war vs. the Soviets, the ASG takes its name from Janjalani's nom de Guerre, Abu Sayyaf, translated as "father of the sword."

9 Maria Ressa, "Investigators Think Sept. 11, 1995 Plot Related," CNN.com, February 25, 2002; Mark Fineman and Richard C. Paddock, "Indonesian Cleric Tied to '95 Anti-U.S. Plot," *Los Angeles Times,* February 7, 2002; Raymond Bonner, "How Qaeda Linked Up with Malaysian Groups," *New York Times,* February 7, 2002.

10 9/11 Commission Staff Statement 15, June 16, 2005 p. 6: "A number of other individuals connected to the 1993 (World Trade Center bombing) and 1995 (Bojinka) plots either were or later became associates of Bin Laden," the Staff wrote in its report, which went on to assert that "We have no conclusive evidence, however, that at the time of the plots any of them was operating under bin Laden's direction."

11 FBI 302 interrogation of Abdul Hakim Murad April 12–13, 1995.

12 Joint Inquiry Statement, September 18, 2002.

13 9/11 Commission Staff Statement 9 April 13, 2004.

14 Matthew Brezinski "Bust and Boom," *Washington Post,* December 30, 2001. The article reported that the CIA official "dismissed the connection to Bojinka as a 'hindsight is cheap' theory."

15 Maria Ressa, "U.S. Warned in 1995 of plot to hijack planes, attack buildings," CNN.com, September 18, 2001; transcript of interview with Secretary Rigoberto Bobby Tiglao, by *CNN,* September 18, 2001.

16 "The Man Who Knew" *Frontline,* http://www.pbs.org/wgbh/pages/frontline/shows /knew/etc/cron.html.

17 Author's interview with Col. Rodolfo B. Mendoza, April 19, 2002.

18 9/11 Commission Staff Statement 16, June 16, 2004.

19 FBI 302 interrogation of Abdul Hakim Murad by Special Agents Frank Pellegrino and Thomas Donlon, April 13, 1995.

20 Peg Tyre "An Icon Destroyed," *Newsweek,* September 11, 2001. *Note:* Several books have given a different accounts. Lou Michel and Dan Herbeck, *American Terrorist:*

Timothy McVeigh and the Tragedy at Oklahoma City (New York; Avon/ReganBooks, revised paperback edition, 2001), use the wording attributed to William A. Gavin, the ADIC of the FBI's New York office who was quoted by Simon Reeve in *The New Jackals,* pp. 108–109. In an interview with the author on June 15, 2003, the retired ADIC said that he lifted Yousef's blindfold and gestured to the Towers with the line, "They're still standing," Yousef reportedly said, "They wouldn't be, if I had had enough money and enough explosives." But in *The Cell,* by Miller, Stone, and Mitchell, FBI agent Chuck Stern is cited as the agent who lifted the blindfold and reminded Yousef that the Trade Center was "still standing." In that version Yousef replied "They wouldn't be, if I'd gotten a little more money" (p. 135).

4. "PLAN TO BLOW UP A PLANE"

1 Author's interview with Steven Legon, Yousef's former attorney, January 20, 2003.
2 Letter from Scarpa Jr.'s attorney, Jeremy Orden to Judge Reena Raggi, April 29, 1999; *U.S. v. Gregory Scarpa Jr.,* 94 Cr. 1119 (S-5).
3 Greg Scarpa Jr. Yousef, "kite."
4 Author's interview with Larry Silverman, April 13, 2004.
5 FBI 302 interview with Greg Scarpa Jr.
6 Author's interview with confidential source.
7 Ibid.
8 Greg Scarpa Jr. Yousef, "kite," May 19, 1996; FBI 302.
9 Ibid.
10 Yousef's technique in which he smuggled two nine volt batteries in the heels of his shoes to escape detection by airport metal scanners, predated the notorious "Shoe Bomber," Richard Reid, by 8 years; Michael Elliott, "The Shoe Bomber's World," *Time,* February 16, 2002.
11 Philip Shenon, "9/11 Panel's Report to Offer New Evidence of Iran-Qaeda Ties," *New York Times,* July 18, 2004.
12 Improvised explosive device; commonly called RDX (Royal Demolition explosive, but also cyclonite, or hexogen). (http://en.wikipedia.org/wiki/RDX).
13 FBI 302 interview with Greg Scarpa Jr. re: Ramzi Yousef intelligence, May 5, 1996.
14 Presidential Daily Briefing, "Bin Laden Determine to Strike in U.S." August 6, 2001, declassified and approved for release April 10, 2004.
15 9/11 Commission, Testimony of Condoleezza Rice, April 8, 2004.
16 Presidential Daily Briefing, August 6, 2001.
17 FBI 302 interview with Greg Scarpa Jr. re: Ramzi Yousef intelligence May 9, 1996.
18 The address of the fictitious front company "Roma Corp." was given to Yousef via Scarpa Jr. as 175 Fifth Avenue, Suite 2252. That happens to the be the address of the Flatiron Building, one of New York's most famous landmarks, located at the intersection of Broadway, Fifth Avenue and Twenty-third Street in Manhattan. The building is occupied today by St. Martin's Press, a publishing company. Curiously, it only has twenty-one floors. There was no twenty-second floor and thus no Suite 2252. This is a fact that would have been easy for one of Yousef's cell members on the outside to check, but for unexplained reasons the FBI located the front company at this address.

19 Greg Smith, "Terrorist Called Pals on Feds' Line," *New York Daily News,* September 24, 2000; Greg Smith, "FBI's Chilling Terror Leads on '96 Tapes," *New York Daily News,* January 21, 2002.
20 Terry McDermott, Jose Meyer, and Patrick J. McDonnell, "The Plots and Designs of Al Qaeda's Engineer," *Los Angeles Times,* December 22, 2002.
21 Joint Inquiry declassified report, July 24, 2003, p. 311.
22 Philippines National Police declassified documents relating to the search of Room 603 at the Dona Josefa Apartments in Manila on January 6, 1995, obtained by the author.
23 FBI 302 interview with Greg Scarpa Jr. re: Ramzi Yousef intelligence May 13, 1996. In this case it was noted that when Yousef learned that the "counselor's phones were down" and Yousef would have to make the call from the tier that morning, he declined.
24 FBI 302 interview with Greg Scarpa Jr. re: Ramzi Yousef intelligence, May 16, 1996.
25 FBI 302 interview with Greg Scarpa Jr. re: Ramzi Yousef intelligence, May 28, 1996.
26 FBI 302 interview with Greg Scarpa Jr. re: Ramzi Yousef intelligence, June 13, 1996.
27 FBI 302 interview with Greg Scarpa Jr. re: Ramzi Yousef intelligence, June 24, 1996.
28 FBI 302 interview with Greg Scarpa Jr. re: Ramzi Yousef intelligence, July 2nd, 1996.
29 Dan Eggen, "9/11 Panel Links Al Qaeda, Iran," *Washington Post,* June 26, 2004.
30 Pat Milton, *In the Blink of an Eye: The FBI Investigation of TWA Flight 800* (New York; Random House, 1999), p. 51.

5. TWA 800: BOJINKA FULFILLED

1 Milton, *In the Blink of an Eye,* p. 3.
2 Ibid., p. 21.
3 *U.S. v. Ramzi Ahmed Yousef et al.,* July 18, 1996.
4 Christopher S. Wren, "Judge to Ask Bomb Trial Jury about Prejudice from Crash." *New York Times,* July 23, 1996.
5 Christopher S. Wren, "The Fate of Flight 800: Legal Fall Out," *New York Times,* July 23, 1996.
6 David Johnston, "The Crash of Flight 800: The Possibilities; Tips, Leads and Theories Are Flooding In," *New York Times,* July 21, 1996.
7 Ibid.
8 Patricia Hurtado, "Bomb Suspect's Letter Revealed," *Newsday,* July 10, 1998.
9 Scarpa-Yousef 302, July 24, 1996.
10 Milton, *In the Blink of an Eye,* p. 51.
11 Matthew L. Wald, "800 Flew for 24 Seconds after the Initial Catastrophe," *New York Times* July 27, 1996.
12 Author's interview with James K. Kallstrom, FBI, ADIC, NYO (ret.), July 12–13, 2004.
13 National Transportation Safety Board, "Witness Group Chairman's Factual Report," TWA Flight 800 Office of Aviation Safety Office of Research and Engineering Washington, D.C. 20594, February 9, 2000. The order to investigator Magladry came from AUSA Caproni during the NTSB's evening progress meeting in the presence of Lewis D. Schiliro, FBI Asst. Director in Charge of the New York office.
14 Author's interview with Larry Silverman, April 13, 2004.
15 Craig Gordon, Earl Lane, and Knut Royce, "PETN: What It Might Mean," *Newsday,* August 24, 1996.

16 Matthew L. Wald, "Fate of Flight 800: The Overview—Jet's Landing Gear Is Said to Provide Evidence of Bomb," *New York Times,* July 31, 1996.
17 Don Van Natta Jr., "More Traces of Explosive In Flight 800," *New York Times,* August 31, 1996.
18 Silvia Adcock and Knut Royce, "Two Traces Found," *Newsday,* August 31, 1996.
19 *U.S. v. Ramzi Yousef et al.*, Testimony of Steven Burmeister, August 15, 1996, p. [illegible] of the court transcript.
20 Liam Pleven and Al Baker, "Probe Follows Chemical Trail," *Newsday,* August 27, 1996.
21 Don Van Natta Jr., "Area Near Wings of 747 Is New Focus of Investigators," *New York Times,* August 13, 1996.
22 Earl Lane, "Support For Tank Explosion Theory," *Newsday,* September 21, 1996.
23 Milton, *In the Blink of an Eye,* pp. 182–183.
24 Author's interview with Supervisory Special Agent Kenneth Maxwell, ret. March, 29, 2003.
25 Louis M. Freeh "Report to the American People on the Work of the FBI 1993–1998."
26 Author's interview with Supervisory Special Agent Ken Maxwell (ret.), July 13, 2004.
27 Don Van Natta Jr., "Prime Evidence Found That Device Exploded in Cabin of Flight 800," *New York Times,* August 23, 1996.
28 Sylvia Adcock and Al Baker, "Probe Turns to Athens for Answers," *Newsday,* September 16, 1996.
29 Michael Arena and Silvia Adcock, "Probers Kept in the Dark," *Newsday,* August 24, 1996.
30 Milton, *In the Blink of an Eye,* p. 227.
31 Ibid., p. 228.
32 Author's interview with confidential FBI source.
33 Dan Barry, "F.B.I. Says 2 Labs Found Traces of Explosive on T.W.A. Jetliner," *New York Times,* August 24, 1996.
34 CNN.com, August 23, 1996.
35 Liam Pleven and Al Baker, "Probe Follows Chemical Trail," *Newsday* August 27, 1996.
36 "Ill Fated TWA Plane Used for Troop Transport in the Gulf War," CNN.com, August 26, 1996; Jack Cashill and James Sanders, *First Strike* (Nashville: WNB Books, 2003).
37 Milton, *In the Blink of an Eye,* p. [illegible].
38 Matthew Purdy, "Bob Security Test on Jet May Explain Trace of Explosives," *New York Times,* September 21, 1996.
39 Don Van Natta Jr., "Setback in T.W.A. Crash Inquiry Adds Urgency to the Search for Evidence of a Bomb," *New York Times,* September 22, 1996.
40 Testimony of James Kallstrom before the House Subcommittee on Aviation, July 10, 1997.
41 Richard Clark, *Against All Enemies: Inside America's War on Terror* (New York: Free Press, 2004).
42 Bill Clinton, *My Life* (New York: Knopf, 2004), p. 718.
43 John Miller, Michael Stone, and Chris Mitchell, *The Cell: Inside the 9/11 Plot and Why the FBI and CIA Failed to Stop It* (New York: Hyperion, 2002), p. 170.

44 David Rohde, "November 9–15; FBI Ends Flight 800 Inquiry," *New York Times*, November 16, 1997.

6. SHATTERING THE K-9 THEORY

1 Letter from James Kallstrom to Congressman James Traficante, September 5, 1997.
2 Cashill and Sanders, *First Strike*, p. 72.
3 Author's phone interview and correspondence with Herman Burnett, May 5, 2004.
4 Author's interview with Herman Burnett, July 11, 2004.
5 Both planes were built at Boeing's plant in Everett Washington and commissioned in October 1971. Milton, *In the Blink of an Eye*, p. 48.
6 Author's phone interview with Herman Burnett, May 3, 2004.
7 Author's interview with John Nance, June 11, 2004.
8 Cashill and Sanders, *First Strike*, p. 74–75.
9 Ibid., p. 74.
10 Testimony of James Kallstrom, http://commdocs.house.gov/committees/Trans/hpw105-33.000/hpw105-33_0.htm.
11 Louis J. Freeh, "Report to the American People on the Work of the FBI 1993–1998," http://hatemonitor.csusb.edu/FBI_Terrorism_rpt/1993-98_fbi%20_report/report7.htm.
12 Illustration by Tom Shoemaker can be found as part of a two-part analysis of the K-9 test, http://www.multipull.com/twacasefile/december.html and http://www.multipull.com/twacasefile/january.html.
13 Author's interview with FBI Supervisory Special Agent Ken Maxwell (ret), March 29, 2003.
14 *U.S. v. Michael Sessa*, CR-92-351 September 24, 2001, pp. 19–20.

7. "THE ULTIMATE PERVERSION"

1 Dannen, "The G-Man and the Hit Man."
2 The figure was cited by Judge Jack B. Weinstein in *Orena v. U.S.*, 956 F. Supp. 1071 (EDNY 1997).
3 Author's interview with Alan Futerfas, May 24, 2004.
4 John Connolly, "Who Handled Who?" *New York*, December 2, 1996.
5 Affidavit of Greg Scarpa Jr., July 30, 2002.
6 In a written statement pursuant to the FBI OPR investigation of his relationship with Scarpa Sr., DeVecchio admitted accepting the doll from the hitman but alleged that it was for "the niece" of a friend. He claimed that he offered to pay for it and that "it would have been an insult" to Scarpa Sr. if he had tried to return it, May 5, 1995.
7 Bill Moushey, "Switching Sides," *Pittsburgh Post-Gazette*, December 1, 1998.
8 *Pasquale Amato and Victor Orena v. the U.S.*, testimony of Gregory Scarpa Jr., January 7, 2004.
9 Moushey, "Switching Sides."
10 Dannen, "The G-Man and the Hit Man."
11 Author's interview with confidential NYPD source.
12 Villano with Astor, *Brick Agent*.
13 Gregory Scarpa Jr., sworn affidavit, March 16, 2003.
14 Futerfas interview, May 24, 2004.

15 *U.S. v. Victor J. Orena,* 92 CR 351 Alan Futerfas and Ellen Resnick; "Final Submission in Support of Victor J. Orena and Pasquale Amato's Motion for a New Trial," filed October 16, 1996, before Judge Jack B. Weinstein, U.S. Court for the Eastern District of New York.
16 Dannen, "The G-Man and the Hit Man."
17 Connolly, "Who Handled Who?"
18 FBI 302 interview with Valerie Caproni, January 26, 1994.
19 FBI 302 interview with Special Agent Christopher Favo, November 16, 1995.
20 Ibid.
21 Ibid.
22 Testimony of R. Lindley DeVecchio, Orena-Amato 2255 hearing, February 27, 1997.
23 Ibid., pp. 137–138.
24 Dannen, "The G-Man and the Hit Man."
25 Ibid.
26 Ibid.
27 Ibid.
28 Alan Futerfas, letter to Judge Charles P. Sifton, February 7, 1995, p. 6.
29 Futerfas interview May 24, 2004.
30 In a later trial, Victor M. Orena Jr., son of the Colombo's boss, and six co-defendants were acquitted based on the defense that they were merely defending themselves against Scarpa Sr.'s murderous attacks.
31 Greg B. Smith, "7 Cleared in Brooklyn Mob Case," *New York Daily News,* July 1, 1995.
32 Author's interview with Flora Edwards, April 27, 2004.
33 Affidavit of Greg Scarpa Jr., 2002.
34 *Pasquale Amato and Victor Orena v. U.S.,* CV-96-1461; CV-96-1474 transcript of hearing before Jack B. Weinstein, United States District Court, Eastern District of New York, January 7, 2004.
35 Ibid. Testimony of Greg Scarpa Jr., January 7, 2004.
36 John Marzulli, "Mobsters Retrial Nixed," *New York Daily News,* January 16, 2004.
37 *U.S. v. Michael Sessa,* CR-92-351, September 24, 2001.
38 Affirmation of Valerie Caproni, May 28, 1996, *U.S. v. Anthony Russo et al.,* 92 CR351 (S9) (CPS).
39 FBI 302 on phone call between Valerie Caproni by SA Thomas Fuentes, January 21, 1994.
40 FBI 302 on phone call between Valerie Caproni by SA Thomas Fuentes and John L. Barrett, February 1, 1994.
41 Dannen, "The G-Man and the Hit Man."

8. THE FORTY-YEAR REWARD

1 Douglas Grover letter to Louis Freeh, May 17, 1995.
2 Jerry Capeci, "My Father Did It," *New York Daily News,* September 28, 1978.
3 FBI 302 memo testimony of Christopher Favo, June 1–2, 1994.
4 Jerry Capeci, "I Spy," *New York Daily News,* October 19, 1998.
5 *U.S. v. Ajaj,* 92-CR-993, EDNY, transcript of phone conversation with Ramzi Yousef aka Rashed.

6 *U.S. v. Ajaj,* 92-CR-993, EDNY, Addendum to the Pre-Sentence Report, December 8, 1992.
7 *U.S. v. Mohammed Ajaj,* transcript of Motion Hearing before Judge Reena Raggi, December 22, 1995.
8 U.S. Attorney for the EDNY letter to Judge Reena Raggi, December 18, 1992.
9 *U.S. v. Salameh,* U.S. Court of Appeals for the Second Circuit, August term.
10 Helen Peterson, "Jury Convicts Mobsters Who Spies in WTC Case," *New York Daily News,* October 24, 1998.
11 *U.S. v. Osama bin Laden et al.,* February 6, 2001.
12 Greg Jr. had been indicted for four murders, but his ultimate conviction was for six counts of racketeering, loan sharking, illegal gambling, and tax fraud. He was acquitted at the trial of five gangland murders, but federal prosecutors succeeded in tying him to four murder conspiracies. Nonetheless, considering that the Feds never linked him personally to a homicide, the four-decade term was particularly harsh, especially since he was ordered to spend it in solitary confinement.
13 Scarpa-Yousef, FBI 302, September 19, 1996.
14 Greg B. Smith, "Terrorist Called Pals on Feds Line," *New York Daily News,* September 24, 2000.
15 Author's interview with Angela Clemente, April 27, 2004.

9. AN NYPD COP TAKES THE FALL

1 Author's interview with Angela Clemente, April 28, 2004.
2 Author's interview with Det. Joseph Simone, April 30, 2004.
3 Carolyn Rushefsky, "Gratitude to Cops," *New York Daily News,* January 24, 1980.
4 Carolyn Rushefsky, "4 Cops Catch 36-Time Rapist and 3 Communities Breathe a Sigh of Relief," *Kings Courier,* October 5, 1981.
5 Carolyn Rushefky, "Burglaries," King's Courier.
6 Glenn Chapman, "Cops Arrest Leader of Alleged Drug Ring," *Staten Island Advance,* January 30, 1985.
7 Dean Balsamini and Tom Berman, "3 Grabbed in Mob Sweep," *Staten Island Advance,* December 31, 1991.
8. Jerry Capeci and Tom Robbins, "Sold Information to Colombos: FBI," *New York Daily News,* December 9, 1993.
9 Author's interview with Det. Joseph Simone, April 27, 2004.
10 *U.S. v. Joseph Simone,* testimony of Christopher Favo, October 18, 1994.
11 Transcript of taped conversation between Salvatore Micciotta and Armando "Chips" DeCostanza, recorded July 14, 1993, 194C-NY-240371.
12 Jerry Capeci and Tom Robbins, "Detective Stung by Feds," *New York Daily News,* December 9, 1993.
13 Jerry Capeci and Tom Robbins, "Mob Biggie Aids FBI Sting," *New York Daily News,* December 11, 1993.
14 Jerry Capeci, "Short FBI Tape May Aid 'Rogue Cop' Defense," *New York Daily News,* April 13, 1994.
15 Greg Smith, "DA Gilds Case for Pet Canary," *New York Daily News,* October 21, 1994.

16 *U.S. v. Joseph Simone,* transcript of cross examination of Special Agent Lynn Smith, pp. 665–669.
17 Author's interview with John Patton, April 29, 2004.
18 FBI 302 interview with Christopher Favo, June 1–2, 1994; November 16, 1995; December 8, 1995.

10. THE DEATH OF NICKY BLACK

1 Joseph P. Fried, "Detective Is Found Not Guilty of Selling Secrets to the Mafia," *New York Times,* October 21, 1994.
2 Pete Bowles, "Cop Not Guilty of Fed Rap," *Newsday,* October 21, 1994.
3 Ann Marie Calzolari, "Detective Acquitted of Mob Charges," *Staten Island Advance,* October 22, 1994.
4 Angela Mosconi, "Cop Puts Life Back Together," *Staten Island Advance* November 6, 1994.
5 Jerry Capeci, "Cop Still Treading Hot Water," *New York Daily News,* November 8, 1994.
6 Fogel Draft, "In the Matter of the Charges and Specifications—against—Detective Joseph Simone," Tax Registry No. 87061, Medical Division before Rae Downes Koshetz, Deputy Commissioner—Trials.
7 Letter from John Patton to NYPD Commissioner Howard Safir, April 26, 1996.
8 Letter from John Patton to NYPD Commissioner Howard Safir, April 26, 1996.
9 Investigators Daily Activity Report, Det. Joseph Simone, January 5, 1992.
10 Author's interview with Flora Edwards, April 27, 2004.
11 Author's interview with Det. Joseph Simone, April 30, 2004.
12 *U.S. v. Gregory Scarpa, Sr.* CR 93 124, May 6, 1993.
13 Dr. Stephen Dresch, e-mail, June 1, 2003.
14 Dr. Stephen Dresch and Angela Clemente, "Pre-September 2001," FBI and Dept. of Justice Intelligence on Terrorism (R. Yousef) sent to Thomas H. Kean, Chairman, and Lee H. Hamilton, Vice Chairman, National Commission on Terrorist Attacks upon the United States, April 5, 2004.

PART II

11. THE WHITE HOUSE STONEWALL

1 Sheryl Gay Stolberg, "9/11 Widows Skillfully Applies the Power of a Question: Why?" *New York Times,* April 1, 2004.
2 Ibid., Kristen Breitweiser, 33, and Patty Casazza, 43, voted Republican; Mindy Kleinberg, 42, and Lorie van Auken, 49, voted Democratic.
3 Andrew Jacobs, "Trade Centers Widows Lobby for Independent Inquiry," *New York Times,* June 11, 2002.
4 "Bush Asks Daschle to Limit Sept. 11 Probe," CNN.com, January 29, 2002.
5 http://www.9-11commission.gov/about/president.htm.
6 Author's interview with Sally Regenhard, November 19, 2003.
7 Linton Weeks, "An Indelible Day," *Washington Post,* June 16, 2004.
8 Author's interview with Lorie van Auken, October 26, 2003.
9 Their official staff bios, http://www.9-11commission.gov/about/bios.htm.

10 Timothy J. Burger, "9-11 Commission Funding Woes," *Time,* March 26, 2003; Editorial, "Undercutting the 9/11 Inquiry," *New York Times,* March 31, 2003.
11 Author's interview with Kyle Hence, September 11, 2003.
12 Scott J. Paltrow, "White House Hurdles Delay 9/11 Commission Investigation," *Wall Street Journal,* July 8, 2003.
13 Author's interview with Beverly Eckert, February 24, 2003.
14 9/11 Commission, First Interim Report, July 8, 2003.
15 Author's interview with confidential Commission staff source.
16 Condoleezza Rice and Philip Zelikow, *Germany Unified and Europe Transformed: A Study in Statecraft* (Cambridge, NJ: Harvard University Press, 1995).
17 Shaun Waterman, "Families: 9/11 Probe Director Has Conflict," United Press International, October 4, 2003.
18 Dan Eggen, "Sept. 11 Panel Defends Director's Impartiality," *Washington Post,* October 14, 2003.
19 Kenneth R. Bazinet, "Bush to 9/11 probers: Briefing Info 'Sensitive,'" *New York Daily News,* October 28, 2003.
20 Shaun Waterman, "9/11 Panel Gets Docs Deal with White House," UPI, November 12, 2003.
21 Laurie Mylroie, in a later footnote we'll cite the *Washington* magazine analysis by Peter Bergen in the October 2003 issue that debunked Mylroie's theory of a link between Saddam, al Qaeda, and 9/11 prior to the invasion.
22 Author's interview with confidential 9/11 Commission staff source, November 8, 2003.
23 Family Steering Committee Statement Regarding the One-Year Anniversary of the National Commission on Terrorist Attacks upon the United States, November 27, 2003.
24 Dan Eggen, "9/11 Panel May Seek Extension," *Washington Post,* November 27, 2003.
25 Boehlert, "The President Ought to Be Ashamed."
26 Philip Shenon, "Ex-Senator Will Soon Quite 9/11 Panel, Leaving Gap for Victim's Advocates." *New York Times,* December 5, 2003.

12. THE CHICKEN COOP AND THE FOX

1 Tom Dunkel, "Restless Mind," *Baltimore Sun,* May 17, 2004.
2 Andrea Alexander, "Widows Seeking Answers on 9/11,"*Asbury Park Press,* December 8, 2003.
3 Randall Pinkston, "9/11 Chair: Attacks Was Preventable," CBS News.com, December 17, 2003.
4 Eric Boehlert, "What Did Bush Know and When Did He Know It?" *Salon.com,* December 19, 2003.
5 Michael Isikoff, "September 11: Will Terror Panel's Report Be an Election Issue?" *Newsweek,* January 12, 2004.
6 Eric Lichtblau and James Risen, "2 on 9/11 Panel Are Questioned on Earlier Security Roles," *New York Times,* January 15, 2004.
7 Philip Shenon and David E. Sanger, "Bush Aides Block Clinton's Papers From 9/11 Panel," *New York Times,* April 2, 2004.
8 Greg Miller, "Panel Investigating 9/11 Attacks Requests More Time to Finish Study," *Los Angeles Times,* January 14, 2004.

9 Dan Eggen, "9/11 Panel Unlikely to Get Later Deadline," *Washington Post,* January 19, 2004.

10 The Family Steering Committee for the 9/11 Independent Commission Statement Regarding the Need for an Extension to January 10, 2005, February 3, 2004.

11 "Bush to Agree to Extend 9/11 Panel Deadline," Reuters, February 4, 2004.

12 Hope Yen, "Sept. 11th Panel Says It Will Pare Down Intelligence Probe If It doesn't Get Extension," Associated Press, February 13, 2004.

13 Robert Cohen, "Hastert Drops Opposition to 9/11 Panel's Request for More Time," *Newark Star Ledger,* February 20, 2004.

14 Hope Yen, "9/11 Panel to Interview Security Advisor," Associated Press, February 7, 2004.

15 Michael Isikoff, "The White House: A New Fight Over Secret 9/11 Docs," *Newsweek,* February 16, 2004.

16 Michael Isikoff and Mark Hosenball, "A recalcitrant Bush administration forces the 9/11 Commission to poke into Bob Woodward's notebook," *Newsweek,* February 18, 2004.

17 Lesley Stahl, "9/11 Before and After," *60 Minutes,* March 21, 2004.

18 Philip Shenon and David E. Sanger, "Bush and Cheney Tell 9/11 Panel of '01 Warnings," *New York Times,* April 30, 2004.

19 Brian Blomquist and Vincent Morris, "Oval Office Insult: Two 9/11 Dems Walk Out on Bush," *New York Post,* April 30, 2004. Former senator Bob Kerry said that he had a previous meeting set up to lobby Sen. Pete Domenici (R-NM) for increased funding for the New School University; Commission co-Chairman Lee Hamilton left to introduce the prime minister of Canada at the Woodrow Wilson Center, where Hamilton is director.

13. YEAR ONE DOGS AND PONIES

1 Philip Shenon, "Panel Plans to Document the Breadth of Lost Opportunities," *New York Times,* April 11, 2004.

2 Jennifer Lee and Eric Lichtblau, "Quiet but Aggressive Staff at Center of 9/11 Inquiry," *New York Times,* April 30, 2004.

3 Philip Shenon, "9/11 Panel May Not Reach Unanimity on Final Report," *New York Times,* May 26, 2004.

4 Statement of Brian Jenkins to the National Commission on Terrorist Attacks Upon the United States, March 31, 2003.

5 Statement of Lee S. Wolosky of Boies, Schiller & Flexner, LLP, to the National Commission on Terrorist Attacks Upon the United States, April 1, 2003.

6 Statement of Bob Graham to the National Commission on Terrorist Attacks Upon the United States May 22, 2003.

7 The tapes were first disclosed in a series of stories by *New York Times* reporter Ralph Blumenthal in the late summer and fall of 1993: "Informer's Ex-Wife Said He Warned of Terrorism," August 28, 1993; "Tapes Depict Proposal to Thwart Bomb Used in Trade Center Blast," October 28, 1993.

8 A copy of the memo is contained as an Appendix on p. 519 in Lance, *1000 Years for Revenge.*

9 Laurie Mylroie, *The War Against America* (New York: ReganBooks/HarperCollins, 2001).
10 Lance, *1000 Years for Revenge,* pp. 161–162.
11 Peter Bergen, *Holy War Inc: Inside the Secret World of Osama bin Laden* (New York: Free Press, 2002), and adjunct professor at the School of Advanced International Studies at Johns Hopkins University.
12 Peter Bergen, "Armchair Provocateur," *Washington Monthly,* December 2003.
13 Statement of Laurie Mylroie to the National Commission on Terrorist Attacks Upon the United States, July 9, 2003.
14 9/11 Commission, Max Cleland; October 14, 2003.
15 Statement of John Deutch to the National Commission on Terrorist Attacks Upon the United States, October 14, 2003.
16 Author's interview with Patty Casazza, July 3, 2004.

14. WARNING: PLANES AS WEAPONS

1 Christopher Marquis, "Reports on Attacks Are Gripping, Not Dry," *New York Times,* June 20, 2004.
2 Within hours of the Commission's pronouncement the president and vice president went on the offensive in the media, insisting that the Iraqi-Al Qaeda ties were solid. David E. Sanger and Robin Toneri, "Bush and Cheney Talk Strongly of Qaeda Links with Hussein," *New York Times,* June 18, 2004. But by July 6 the Commission responded to the vice president, disputing his suggestion that he had access to more intelligence than the Commission did about he alleged Iraqi-Al Qaeda ties. Philip Shenon, "Leaders of 9/11 Panel Answer Information Claim by Cheney," *New York Times,* July 7, 2004.
3 Philip Shenon and Christopher Marquis, "Panel Finds No Qaeda-Iraq Tie: Describes a Wider Plot for 9/11," *New York Times,* June 17, 2004.
4 Philip Shenon and Christopher Marquis, "Panel Says Chaos in Administration Was Wide on 9/11," *New York Times,* June 18, 2004.
5 Douglas Jehl, "The Reach of War: Intelligence—No Saudi Payment to Qaeda Is Found," *New York Times,* June 19, 2004.
6 PNP memo: "Initial Results re: Investigation on Suspected Islamic Extremists," January 6, 1995; PNP Person Report on Mohammed Jamal Khalifa, August 25, 1996. John M. Berger, "INS Deported al Qaeda-Linked Suspect Just Days After Oklahoma Bombing," Intelwire.com. Berger's online article ended by noting that Khalifa claims to be a philanthropist and businessman and he has repeatedly and vehemently denied any involvement in terrorist activities, most recently in a September 2003 interview with *Arab News.* "'I spend all my life helping people,' Khalifa tearfully told his immigration judge in April 1995." On April 12, Berger reported that Khalifa, one of 200 mostly Saudi defendants in a lawsuit brought on behalf of 9/11 victims' families by the law firm of Motley Rice, sought to withdraw from the action: "Bin Laden Brother-in-Law Khalifa Moves to be Dropped from 9/11 Suit," Intelwire.com, April 12, 2004.
7 9/11 Commission Staff Statement No. 1, January 26, 2004.
8 9/11 Commission Staff Statement No. 2, January 26, 2004.
9 9/11 Commission Staff Statement No. 3, January 26, 2004.

10 "U.S.: Al Qaeda Is 70 Percent Gone, Their 'Days Are Numbered," Worldtribune.com, January 23, 2004.
11 Final Declassified Report of the Joint Inquiry, July 24, 2003, pp. 209–215.
12 9/11 Commission Staff Statement No. 4, January 26, 2004.
13 Ibid.
14 Ibid.
15 Accused CIA killer Mir Aimal Kansi, aka Kasi.
16 *U.S. v. Ramzi Yousef et al.*, August 9, 1996. In testimony at the Bojinka trial it is established than an FBI LEGAT (legal attaché) from Bangkok who was in Islamabad at the time of the Yousef takedown got to the Guest House twenty minutes after Yousef's arrest and that Special Agent Brad Garrett arrived five minutes after him: Samuel M. Katz, *Relentless Pursuit: The DSS and the Manhunt for the Al-Qaeda Terrorists* (New York: Forge, 2002).
17 Christopher John Farley, "The Man Who Wasn't There," *Time,* May 20, 1995.
18 "How the FBI Gets Its Man," *60 Minutes II,* October 10, 2001.
19 Bill Gertz, *Breakdown: How America's Intelligence Failures Led to September 11* (Washington, D.C.: Regnery, 2002).
20 Brian Ross and David Scott, "An Al Qaeda Ally? Member of Qatari Royal Family Helped Senior Al Qaeda Official Get Away," *ABC News,* February 7, 2003.
21 Josh Meyer and John Goetz, "War with Iraq: Qatar's Security Chief Suspected of Having Ties to Al Qaeda; Interior Minister Reportedly Sheltered Terrorists, Including 9/11 Plotter. U.S. Military Campaign's Headquarters Are in the Gulf Nation's Capital," *Los Angeles Times,* March 28, 2003.
22 Joint Inquiry Report, September 23, 2002.
23 Ibid.
24 Meyer and Goetz, "War with Iraq."

15. "ALARMING THREATS" POURING IN

1 9/11 Commission Staff Statement No. 6, March 23, 2004.
2 9/11 Commission Staff Statement No. 7, March 24, 2004.
3 9/11 Commission Staff Statement No. 8, March 24, 2004.
4 Mimi Hall, "Poll Finds Most Are Put Off by Bush Vacation," *USA Today,* August 6, 2001.
5 Dan Priest, "Tenet: System Was Blinking Red." *Washington Post,* April 15, 2004.
6 Elizabeth Bumiller and Douglas Jehl, "Tenet Resigns as C.I.A. Director: 3 Harsh Reports on Agency Due," *New York Times,* June 4, 2003.
7 Scott Lindlaw, "Rice Rejecting Calls for Public Testimony," Associated Press, March 28, 2004.
8 Ed Bradley, "Rice Leads Counterattack," *60 Minutes,* March 28, 2004.
9 "9/11 Commission Demands Rice Testify," U.P.I., March 30, 2004.
10 Philip Shenon and Eric Schmitt, "Bush and Clinton Aides Grilled by Panel," *New York Times,* March 24, 2004.
11 "*Newsweek* Poll: The 9/11 Commission," *Newsweek,* March 27, 2004.
12 Philip Shenon and Elizabeth Bumiller, "Bush Allows Rice to Testify on 9/11 in a Public Session," *New York Times,* March 31, 2004.

13 Elizabeth Bumiller, "The Presidents' Credibility Now Rests on Her Shoulders," *New York Times,* April 4, 2004.
14 "A Stream of Threats, an Emerging Strategy and an Unseen Plot," *New York Times,* April 4, 2004.
15 Elizabeth Bumiller and Philip Shenon, "No Public Appearance by the Person Panel Wants Most," *New York Times,* March 26, 2004.
16 Nicholas Lemann, "Without a Doubt," *New Yorker,* October 10, 2003.
17 White House press briefing, May 17, 2002, CNN.com.
18 9/11 Commission, testimony of Condoleezza Rice, April 8, 2004.

16. CHECKING THE BUREAU SPIN

1 Lance, *1000 Years for Revenge,* pp. 93–94.
2 Ibid., pp. 101, 112, 141–142, 185, 202–203.
3 Ibid., pp. 291–293.
4 Ibid., pp. 136–137, 150–152. Note one change: Where previously Salem had refused to wear a wire and testify in open court, he agreed to do so for this second investigation. In *1000 Years for Revenge* we discussed at length the FBI's options in the spring of 1992, when Dunbar, insisting that Salem wear a wire, could have used alternate investigative methods to achieve probable cause for indictments and convictions in what was known as the "twelve Jewish locations" plot. Surveillance could have been conducted of the cell tied to El Sayyid Nosair, Title III wiretap warrants could have been obtained to monitor their phone calls, and another FBI informant could have been inserted into the cell to corroborate Salem's findings at trial. But Dunbar chose none of these options. In fact, after Salem had passed one polygraph and furnished Agent Floyd with reliable evidence for months, the ASAC demanded that Salem submit to two more lie detector tests before he ultimately withdrew. Once Salem was out of the picture, Sheikh Rahman contacted Pakistan, Ramzi Yousef flew to New York, and the "twelve Jewish locations" plot escalated into the World Trade Center bombing conspiracy.
5 Ibid., pp. 145–148.
6 Ibid., pp. 155–156.
7 Ibid., pp. 169–173, 222, 245–247, 376–378.
8 Ibid., p. 430.
9 Ibid., pp. 373–374.
10 Craig Unger, "Saving the Saudis," *Vanity Fair,* October 2003; *House of Bush, House of Saud* (New York: Scribner, 2004); "The Great Escape," *New York Times,* June 1, 2004.
11 Kathy Steele, "Phantom Flight from Florida," *Tampa Tribune,* October 5, 2001.
12 Jean Heller, "TIA Now Verifies Flight of Saudis," *St. Petersburg Times,* June 9, 2004.

17. "NOT A SINGLE PIECE OF PAPER"

1 9/11 Commission Staff Statement No. 11, April 14, 2004.
2 *U.S. v. Osama bin Laden,* February 6, 2001.
3 9/11 Commission, testimony of Special Agent Mary Deborah (Debbie) Doran, June 16, 2004.

4 9/11 Commission Final Report, PDF p. 62, July 22, 2004.

5 John Berger, "INS Deported al Qaeda-Linked Suspect Just Days After Oklahoma City Bombing," Intelwire.com.

6 John Berger, "FBI Informant Tipped Authorities about Airplane Crash Plot in 1993," Intelwire.com, April 8, 2004.

7 Final Declassification Report of the Joint Inquiry, July 24, 2003, pp. 209–213.

8 Remarks prepared for delivery by Robert S. Mueller at the Commonwealth Club, April 19, 2002.

9 David Johnston, "FBI Says Pre-Sept 11 Call for Inquiry Got Little Notice," *New York Times,* May 9, 2002; David Johnston, "FBI Director's Comments," *New York Times,* May 31, 2002; Statement for the Record of Robert S. Mueller, III, Director Federal Bureau of Investigation, on FBI Reorganization Before the Senate Committee on the Judiciary, May 8, 2002. *Note:* The remarks quoted were not contained in Mueller's prepared text but came during questioning by the Committee members.

10 Letter from Coleen Rowley, FBI Minneapolis, to Robert S. Mueller, FBI director, May 21, 2002.

11 Mueller made good on this goal and went beyond it, establishing a total of eighty-four JTTFs by April 2004.

12 Robert S. Mueller, "Reforming the FBI in the 21st Century," testimony before the Senate Judiciary Committee, May 8, 2002.

13 Robert S. Mueller, "Progress Report on the Reorganization and Refocus of the FBI," testimony before the House Appropriations Subcommittee, June 18, 2003.

14 9/11 Commission Staff Statement No. 9, April 13, 2004.

15 "Washington in Brief," *Washington Post,* December 11, 2003.

16 "A Review of the Fbi's Trilogy Information Technology Modernization Program," National Research Council, May 10, 2004.

17 John Schwartz and Lowell Bergman, "FBI Sees Delay in New Network to Oversee Cases," *New York Times,* June 26, 2004.

18 James Ridgeway, "This Made Ashcroft Gag," *Village Voice,* May 24, 2004.

19 Eric Lichtblau, "Material Given to Congress in 2002 Is Now Classified," *New York Times,* May 20, 2004.

20 Andrew Buncombe, "Whistleblower the White House Wants to Silence, Speaks to the Independent," *Independent,* April 2, 2004; Andrew Buncombe, "Leak: U.S. knew of Sept 11 plans," *Washington Times*/United Press International, April 2, 2004.

21 9/11 Commission hearing, April 14, 2004, testimony of Richard Ben-Veniste

22 *The 9/11 Comission Report* (New York; WW. Norton, 2004), chapter 3, footnote 25.

23 "Briton Trained as Suicide Hijacker," *Sunday Times of London,* May 9, 2004.

24 Sue Reid, "He's an Ex-Waiter and Gambler from Oldham Who Claims He Trained as a Terrorist," *London Daily Mail,* May 25, 2004.

25 Lisa Myers, "Did al-Qaeda Trainee Warn FBI before 9/11?" *NBC Nightly News with Tom Brokaw,* June 3, 2004.

26 FBI 302 interview with Greg Scarpa Jr., September 9, 1996.

27 FBI 302 interview with Greg Scarpa Jr., December 26, 1996.

28 Ibid.

29 Siobhan McDonough, "Kin Upset Hijacker-in-Training Freed," Associated Press, June 3, 2004.

30 Philip Shenon and Kevin Flynn, "Former City Officials Strongly Rebut Criticism of 9/11 Panel." *New York Times,* May 18, 2004.
31 "Giuliani Describes 9/11 Rescue Efforts," CNN.com, May 20, 2004.

18. THE "LOOSE NETWORK" BEHIND 9/11

1 Phil Shenon, "Ex Senator Kerrey Is Named to Federal 9/11 Commission," *New York Times,* December 10, 2003.
2 Dan Eggen "Kerry Replacing Member of 9/11 Panel," *Washington Post,* December 10, 2003.
3 Gregory L. Victica, "What Happened in Thanh Phong," *New York Times Magazine,* April 29, 2001.
4 Tom Dunkel, "Restless Mind," *Baltimore Sun,* May 17, 2004.
5 Bill Vann, "War Criminal to Probe Mass Murder: Ex-Senator Bob Kerrey Appointed to 9/11 Panel," WSWS.com, December 12, 2003.
6 *U.S. v. Mohammed Salameh et al.,* S5 93 CR.180 (KTD); *U.S. v. Omar Abdel Rahman et al.,* S5 93 CR.181 (MBM); *U.S. v. Ramzi Ahmed Yousef et al.,* S1 293 CR.180 (KTD); *U.S. v. Ramzi Ahmed and Eyad Ismoil,* S1 293 CR.180 (KTD), *U.S. v. Osama bin Laden et al.,* S 7 98 CR.1023.
7 Author's interview with Col. Rodolfo B. Mendoza, April 19, 2002.
8 Quoting directly from the FBI 302 with respect to his backers: "BASIT (YOUSEF) would not elaborate on exactly how the WTC bombing was financed, except to say that he had received money from family and friends." FBI 302, "Interrogation of Ramzi Ahmed Yousef by Chuck Stern & Brian Parr," p. 14. "When questioned regarding a business card in the name of MOHAMMED KHALIFA, found in BASIT'S apartment in the Philippines BASIT stated that he did not personally know KHALIFA, but that KHALIFA's business card had been given to him by WALI SHAH, as a contact in the event BASIT needed aid. BASIT also acknowledged that he was familiar with the name USAMA BIN LADEN, and knew him to be a relative of KHALIFA's but would not further elaborate." Ibid., p. 19.
9 *U.S. v. Sattar et al.,* Si o2 CR.395 (JGK), indictment, November 19, 2002.
10 PNP: "Initial Results re Investigation on Suspected Islamic Extremists," January 6, 1995; PNP Person Report on Mohammed Jamal Khalifa, August 25, 1996.
11 Benjamin Weiser, "The Terror Verdict: The Organization—Trial Poked Holes in Image of bin Laden's Terror Group," *New York Times,* May 31, 2001.
12 James Risen, "A Nation Challenged: Al Qaeda—Bin Laden Aide Reported Killed by U.S. Bombs," *New York Times,* November 17, 2001.
13 "Al Qaeda Claims Tunisia Attack: Nineteen People Died in the Attack on El-Ghriba," bbc.co.uk, June 23, 2003.
14 Joe Katzman, "Breaking News: Karachi Attack," windsofchange.net, June 14, 2002.
15 Laura Hayes, "Al Qaeda, Osama bin Laden's Network of Terror," infoplease.com.
16 "Terror Mastermind Hambali Had Fake Spanish Passport When Arrested," *Jakarta Post*/Associated Press, June 24, 2004.
17 John Diamon, "U.S. Holding Alleged Mastermind of Bali, Jakarta Bombings," *USA Today,* August 15, 2003.
18 "Saudi Ambassador Says 600 Arrested After Terror Bombing," *Pakistan Daily Times,* July 7, 2004.

19 "Terror in Turkey, Al Qaeda Claims Responsibility," india-newsbehindnews.com, November 24, 2003.
20 Gethin Chamberlain, "Mass Murder in Madrid—Al Qaeda Link," *Scotsman,* March 12, 2004.
21 Bruce Stanley, "Saudis Reassure Investors after New Attacks," *Sun Newspapers* [Canada], canoe.com, May 31, 2004.
22 "Letter May Detail Iraqi Insurgency's Concerns," CNN.com, February 10, 2004.
23 9/11 Commission Staff Statement No. 15, June 16, 2004.
24 Benjamin Weiser, "Mastermind Gets Life for Bombing of Trade Center," *New York Times,* January 10, 1998.
25 Author's interview with Col. Mendoza, April 19, 2002.
26 9/11 Commission Staff Statement No. 16, June 16, 2004.
27 David Johnston with Don Van Natta Jr., "After Effects: Terrorist Threat—U.S. Officials See Signs of a Revived Al Qaeda," *New York Times,* May 17, 2003.
28 Author's interview with Col. Mendoza, April 19, 2001.
29 *U.S. v. Ramzi Ahmed Yousef et al.,* July 18, 1996.
30 *U.S. v. Ramzi Ahmed Yousef et al.,* August 6, 1996.
31 FBI 302, "Interrogation of Abdul Hakim Murad," April 12–13, 1995.
32 Scarpa-Yousef 302, July 24, 1996.
33 Peter Lance, *1000 Years for Revenge,* pp. 510–511.
34 Greg B. Smith, "On Clinton's Watch: Feds Nixed Deal for Plane Plot Tipoff," *New York Daily News,* September 25, 2001.

19. EIGHTEEN MINUTES TO CALL NORAD

1 Thompson's research is also available in his book *The Terror Timeline* (New York: ReganBooks/HarperCollins, 2004).
2 William B. Scott, "Exercise Jump-Starts Response to Attacks," *Aviation Week and Space Technology,* June 3, 2002.
3 Ibid.; Mike Kelly, "Atlantic City F-16 Fighters Were Eight Minutes Away from 9/11 Hijacked Planes," *Bergen Record,* December 5, 2003.
4 John J. Lumpkin, "Agency Was to Crash Plane on 9–11," Associated Press, August 22, 2002; Pamela Hess, "U.S. Agencies—Strange 9/11 Coincidence," United Press International, August 22, 2002.
5 Roemer asked General Ralph E. Eberhart of NORAD, "You were prepared for an exercise against the former Soviet Union. Did that help or hurt? Did that help in terms of were more people prepared? Did you have more people ready? Were more fighters fueled with more fuel? Or did this hurt in terms of people thinking, 'No, there's no possibility that this is real world. We're engaged in an exercise,' and delay things?" The General's response: "Sir, my belief is that it helped because of the manning, because of the focus, because the crews—they have to be airborne in fifteen minutes, and that morning, because of the exercise, they were airborne in six or eight minutes. And so, I believe that focus helped." 9/11 Commission, hearing testimony, June 17, 2004.
6 Phil Taylor & Anna Gekoski, "Terror in America: Analysis," *News of the World,* September 16, 2001; Curtis Morgan, David Kidwell, and Oscar Corral, "Prelude to Terror," *Miami Herald,* September 22, 2001.

7 Newsbytes, "Odigo Clarifies Attack Messages," *Washington Post,* September 28, 2001; Dror Yuval, "Odigo Says Workers Were Warned of Attack," *Ha'aretz,* September 26, 2001.

8 "About 8:20," Sylvia Adcock, Bryan Donovan, and Craig Gordon, "America's Ordeal: Where System Failed—Air Attack on Pentagon Indicates Small Eye Weaknesses," *Newsday,* September 23, 2001; "about 8:20," Matthew L. Wald, "Pentagon Tracked Deadly Jet but Found No Way to Stop It," *New York Times,* September 15, 2001.

9 Bryan Ross and Jill Rackmill, "Witnesses to Tragedy: Air Traffic Controllers Haunted by Memories of Sept. 11," *ABC News,* September 6, 2002.

10 9/11 Commission Staff Statement No. 17, June 17, 2004: "Herndon (Va.) prompted the Command Center to notify some FAA field facilities that American 77 was lost."

11 William B. Scott, "Exercise Jump-Starts Response to Attacks," *Aviation Week and Space Technology,* June 3, 2002; "9/11: Interviews by Peter Jennings," *ABC News,* September 11, 2002; Hart Seely, "Amid Crisis Simulation, 'We Were Suddenly No-Kidding Under Attack,'" *Newhouse News,* January 25, 2002.

12 Seely, "Amid Crisis Simulation, 'We Were Suddenly No-Kidding Under Attack,'."

13 Ibid.; 9/11 Commission Report, chapter 1, end notes 134 and 154, p. 461.

14 "9/11: Interviews by Peter Jennings," *ABC News.*

15 A NORAD spokesperson told the *Boston Globe* what typically happens when a fighter scrambles: "When planes are intercepted, they typically are handled with a graduated response. The approaching fighter may rock its wingtips to attract the pilot's attention, or make a pass in front of the aircraft. Eventually, it can fire tracer rounds in the airplane's path, or, under certain circumstances, down it with a missile." Glen Johnson, "Facing Terror Attack's Aftermath: Otis Fighters Scrambled Too Late to Halt the Attacks," *Boston Globe,* September 15, 2001.

16 Kevin Dennehy, "I Thought It Was the Start of World War III," *Cape Cod Times,* October 21, 2002.

17 "Officials: Government Failed to React to FAA Warning," *CNN,* September 17, 2001; 8:53: "Timeline in Terrorist Attacks of Sept. 11, 2001," *Washington Post,* September 12, 2001; Bradley Graham, "Military Alerted before Attacks: Jets Didn't Have Time to Intercept Hijackers Officers Say," *Washington Post,* September 15, 2001; 8:52: "9/11: Interviews by Peter Jennings," *ABC News.*

18 Tech. Sgt. Rick DelaHaya, "F-15 Eagle Celebrates Silver Anniversary," *Air Force News,* July 30, 1997.

19 Richard Whittle, "National Guard Fighters Race After Two Airliners," *Dallas Morning News,* September 16, 2001.

20 "9/11: Interviews by Peter Jennings," *ABC News.*

21 Kevin Dennehy, "I Thought It Was the Start of World War III."

22 As *CNN* reported in 1999, "Only the president has the authority to order a civilian aircraft shot down": Judy Woodruff, Charles Zewe, and Jaime McIntyre, "Investigation into Mysterious Crash of the Famous Payne Stewart Learjet Begins: Unusual Circumstances of Crash Raises Disturbing Question," *World View, CNN,* October 26, 1999.

23 "Officials: Government Failed to React to FAA Warning," *CNN;* 8:43: "Timeline in Terrorist Attacks of Sept. 11, 2001," *Washington Post;* 8:43: Calvin Woodward, "Farther from Sept. 11 Horror, a Refined View of How the Plot Unfolded; NY454, NY456, NY455, NY475, NY494, All of August 12; WX110-WX116 and VACHA501

of Aug. 19," Associated Press, August 19, 2002; 8:43: Sylvia Adcock, "A Loss of Control," *Newsday,* September 10, 2002.

24 Seely, "Amid Crisis Simulation, 'We Were Suddenly No-Kidding Under Attack,'."

25 Calvin Woodward, "Dateline Washington," Associated Press, September 15, 2001; "Chairman Christopher Cox's Statement on Terrorist Attack on America," September 11, 2001.

26 Bill Hemmer, Barbara Starr, "The Pentagon Goes to War: National Military Command Center," *American Morning with Paula Zahn, CNN,* September 4, 2002.

27 Ibid.

28 "September 11: Chronology of Terror," *CNN,* September 12, 2001; "The Tragic Timeline," *New York Times,* September 12, 2001; "Officials: Government Failed to React to FAA Warning," *CNN;* "NORAD's Response Times," NORAD, September 18, 2001; "Timeline in Terrorist Attacks of Sept. 11, 2001," *Washington Post;* Woodward, "Farther from Sept. 11 Horror"; Martha T. Moore and Dennis Cauchon, "Delay Meant Death on 9/11," *USA Today,* September 3, 2002; Marilyn Adams, Alan Levin, and Blake Morrison, "Special Report—No One Sure if Hijackers Were on Board," *USA Today,* August 13, 2002; Adcock, "A Loss of Control"; Michael Ellison, "'We Have Planes, Stay Quiet'—Then Silence," *Guardian,* October 17, 2001; "Events of September 11, 2001," *MSNBC,* September 22, 2001.

29 William B. Scott, *Aviation Week and Space Technology,* September 9, 2002.

30 "Address to the Nation on the Terrorist Attacks," *Public Papers of the Presidents,* September 11, 2001.

31 *MSNBC,* October 29, 2002.

32 "September 11, 2001—A Timeline," *Christian Science Monitor,* September 17, 2001; Johanna McGeary and David Van Biema, "The New Breed of Terrorist," *Time,* September 12, 2001.

33 "9/11: Interviews by Peter Jennings," *ABC News.*

34 "Officials: Government Failed to React to FAA Warning," *CNN;* "NORAD's Response Times," NORAD; "Timeline in Terrorist Attacks of Sept. 11, 2001," *Washington Post;* "A Day of Terror: The Measurement; Columbia's Seismographs Log Quake-Level Impacts," *New York Times,* September 12, 2001; Ellison, "'We Have Planes, Stay Quiet'—Then Silence"; "September 11: Chronology of Terror," *CNN,* September 12, 2001; Woodward, "Farther from Sept. 11 Horror"; Adcock, "A Loss of Control"; Moore and Cauchon, "Delay Meant Death on 9/11"; Adams, Levin, and Morrison, "Special Report—No One Sure if Hijackers Were on Board"; "Events of September 11, 2001," *MSNBC;* Dan Balz and Bob Woodward, "America's Chaotic Road to War," *Washington Post,* January 27, 2002; Elisabeth Bumiller and David E. Sanger, "Threats and Responses: The White House—Threat of Terrorism Is Shaping Focus of Bush Presidency," *New York Times,* September 11, 2002; *USA Today,* December 20, 2001.

35 David E. Sanger and Don Van Natta Jr., "After the Attacks: The Events—In Four Days, a National Crisis Changes Bush's Presidency," *New York Times,* September 16, 2001; 9:05: William Langley, "Revealed: What Really Went on during Bush's 'Missing Hours'," *The Telegraph,* December 16, 2001; 9:05: *Albuquerque Tribune,* September 19, 2002; 9:07: Bill Sammon, "Suddenly a Time to Lead: Difficult Moment for America Transforms the President," *Washington Times,* October 8, 2002.

36 William B. Scott, "F-16 Pilots Considering Ramming Flight 93," *Aviation Week and Space Technology,* September 9, 2002.
37 After 9/11 the "mission statement" on its web site was changed, describing a "vision" to "provide peacetime command and control and administrative mission oversight to support customers in achieving the highest levels of readiness": DCANG Home Page before and after the change.
38 "9/11: Interviews by Peter Jennings," *ABC News.*
39 At one point the estimate of hijacked jets goes as high as eleven.
40 "NORAD's Response Times," NORAD; 9:24: Woodward, "Farther from Sept. 11 Horror"; 9:25: Michael Holmes and Jamie McIntyre, "America's New War: Bush Ordered Flights Shot down If Threatened Capitol," *CNN,* September 17, 2001; 9:25: Glenn Kessler and Don Philips, "Air Travel System Grounded for First Time: Travelers and Carriers Face Day of Chaos, Future Questions," *Washington Post,* September 12, 2001; 9:25: Ellison, " 'We Have Planes, Stay Quiet'—Then Silence."
41 Adcock, Donovan, and Gordon, "America's Ordeal: Where System Failed—Air Attack on Pentagon Indicates Small Eye Weaknesses"; 9:24; "NORAD's Response Times," NORAD; 9:27: Holmes and McIntyre, "America's New War: Bush Ordered Flights Shot Down if Threatened Capitol"; 9:25: Kessler and Philips, "Air Travel System Grounded for First Time: Travelers and Carriers Face Day of Chaos, Future Questions"; 9:35: Holmes and McIntyre, "America's New War: Bush Ordered Flights Shot down If Threatened Capitol"; 9:35; Graham, "Military Alerted before Attacks: Jets Didn't Have Time to Intercept Hijackers, Officers Say."
42 Scot J. Paltrow, "Government Accounts of 9/11 Reveal Gaps, Inconsistencies," *Wall Street Journal,* March, 22, 2004.
43 "NORAD's Response Times," NORAD; 9:37: "Timeline in Terrorist Attacks of Sept. 11, 2001,"*Washington Post;* 9:37: Holmes and McIntyre, "America's New War: Bush Ordered Flights Shot down If Threatened Capitol"; 9:37: Ellison, " 'We Have Planes, Stay Quiet'—Then Silence"; 9:37: Adams, Levin, and Morrison, "Special Report—No One Sure if Hijackers Were on Board"; 9:37: "9/11: Interviews by Peter Jennings," *ABC News;* 9:37: Scott Pelley, "The President's Story: Behind the Scenes with President Bush and His Staff on 9/11," *60 Minutes II,* CBS, September 11, 2002; 9:39: Balz and Woodward, "America's Chaotic Road to War"; 9:40: "Related Major Developments in the Year Since the Sept. 11 Terrorist Attacks," Associated Press, August 19, 2002; 9:43: "Chronology: The Day After," *CNN,* September 12, 2001; 9:43: Karen Breslau, "The Final Moments of United Flight 93," *Newsweek/MSNBC,* September 22, 2001; 9:43: *MSNBC,* September 3, 2002; 9:43: "The Tragic Timeline," *New York Times;* 9:45: Glen Johnson, "Fighting Terror: The Hijackings in the Cockpits—Probe Reconstructs Horror, Calculated Attacks on Planes," *Boston Globe,* November 23, 2001.
44 NORAD September 18, 2001.
45 9/11 Commission Report, June 17, 2004.
46 *Wall Street Journal,* March 22, 2004; 9/11 Commission Report, June 17, 2004.
47 Balz and Woodward, "America's Chaotic Road to War."
48 "Forty Lives, One Destiny; Fighting Back in the Face of Terror," *Pittsburgh Post-Gazette,* October 28, 2001.
49 "Suicide Option: September 11 Details Revealed," *World News Tonight with Peter Jennings, ABC News,* August 30, 2002; "Special and Breaking News 9/11," *ABC News,* September 11, 2002.

50 Balz and Woodward, "America's Chaotic Road to War."
51 Phil Hirschkorn and David Mattingly, "Families Say Flight 93 Tapes Prove Heroism," *CNN,* April 19, 2002; Jere Longman, *Among the Heroes* (New York: HarperCollins, 2002), pp. 270–271.
52 Richard Wallace, "What Did Happen to Flight 93?" *Mirror,* September 13, 2002. The cockpit voice recording of Flight 93 was recorded on a thirty minute reel, which means that the tape was continually overwritten and only the final thirty minutes of any flight would be recorded. The government later permits relatives to hear this tape. Apparently, the version of the tape played to the family members begins at 9:31 A.M. and runs for thirty-one minutes, ending one minute before the plane crashes, according to the government. CNN.com, April 19, 2002; Longman, *Among the Heroes,* pp. 206–7. The *New York Observer* comments, "Some of the relatives are keen to find out why, at the peak of this struggle, the tape suddenly stops recording voices and all that is heard in the last 60 seconds or so is engine noise. Had the tape been tampered with?" *New York Observer,* June 17, 2004.
53 "NORAD's Response Times," NORAD.
54 William Bunch, "Cockpit Voice Recording Ends before Flight 93's Official Time of Impact," *Philadelphia Daily News,* September 16, 2002.
55 "9/11: Interviews by Peter Jennings," *ABC News.*
56 *Washington Post,* March 28, 2004.
57 United Press International, April 10, 2004.
58 Clarke, *Against All Enemies,* pp. 2–4.
59 Author's interview with Monica Gabrielle, April 30, 2004.

20. "AMERICA'S UNDER ATTACK"

1 Shaun Waterman, "9/11 Commission Finishes First Chapters," U.P.I., July 1, 2004.
2 Author's interview with Monica Gabrielle, July 2, 2004.
3 Waterman, "Families: 9/11 Probe Director Has Conflict."
4 Philip Shenon, "9/11 Panel Choose Publisher for Report," *New York Times,* May 25, 2004.
5 Philip D. Zelikow, Timothy Naftali, and Ernest May, *The Presidential Recordings: John F. Kennedy, Volumes 1–3: The Great Crises* (New York: W.W. Norton, 2001).
6 Robert B. Zoellick and Philip D. Zelikow, *America and the East Asian Crisis: Memos to a President* (New York: W.W. Norton, 2000).
7 http://www.ustr.gov/about-ustr/ambassador/zoellick.html.
8 Author's interview with Ron Motley, March 31, 2004.
9 J. Scott Orr, "Retailers Snap Up 9/11 Commission's Still Unpublished Report," *Newark Star Ledger,* June 25, 2004.
10 Philip Shenon, "And Now for the Hard Part," *New York Times,* June 20, 2004.
11 "Roemer: 9/11 Commission Report Likely to Be Shocking," Associated Press, May 7, 2004.
12 Shenon, "And Now for the Hard Part."
13 Sheryl Gay Stolberg, "Panel Members, Insiders All, Question Friends, but Too Gingerly for Some Viewers," *New York Times,* March 25, 2004.
14 Sheryl Gay Stolberg, "Relatives of the Lost Want a Longer, More Open Book," *New York Times,* June 17, 2004.

15 "9/11 Commission's Report to Avoid Ginger Pointing," *Washington Times*/Associated Press June 10, 2004.
16 Author's interview with Mindy Kleinberg, June 11, 2004.
17 William Bunch, "Why Don't We Have Answers to These 9/11 Questions," *Philadelphia Daily News,* September 11, 2003.
18 Boehlert, "The President Ought to Be Ashamed."
19 Marie Cocco, "On 9/11 Underlings Outshone Biggies," *Newsday,* June 22, 2004.
20 Steven Komarow and Tom Squitieri, "NORAD Had Drills Eerily Like Sept. 11," *USA Today,* April 19, 2004.
21 Transcript *Paul Zahn Now,* CNN.com, June 17, 2004.
22 Editorial, "End Panel Theatrics," *USA Today,* May 21, 2004.
23 "Report on the U.S. Intelligence Community's Prewar Intelligence Assessments on Iraq," Senate Select Committee on Intelligence, released July 9, 2004.
24 Douglas Jehl, "Tenet Resigns as C.I.A. Director: 3 Harsh Reports on Agency Due," *New York Times,* June 4, 2004.
25 "CIA Covert Operations Chief Retiring," CNN.com, June 5, 2004.
26 Douglas Jehl, "C.I.A. Director Again Disputes Hijacker's Iraqi Contact," *New York Times,* July 8, 2004.
27 Douglas Jehl, "White House and C.I.A. Withhold Documents," *New York Times,* July 14, 2004.
28 Author's interview with Monica Gabrielle, July 14, 2004.
29 "Patterns of Global Terrorism 2003," U.S. State Department, http://www.state.gov/s/ct/rls/pgtrpt/2003.
30 Dan Eggen, "Powell Calls Report 'A Big Mistake'," *Washington Post,* June 14, 2004.
31 "U.S.: Al Qaeda Is 70 Percent Gone, Their 'Days Are Numbered,'" worldtribune.com, January 23, 2004.
32 R. Jeffrey Smith, "State Dept. Concedes Errors in Terror Data," *Washington Post,* June 10, 2004.
33 Eggen, Ibid.
34 "U.S. Wrongly Reported Drop In World Terrorism in 2003," Associated Press/*New York Times,* June 11, 2003.
35 Author's interview with Lorie van Auken, July 14, 2004.
36 Eric Lichtblau, "Report Questions the Value of Color Coded Warnings," *New York Times,* July 13, 2004.
37 "GAO Report Criticizes Terror Warnings," CNN.com, July 13, 2004.
38 Lichtblau, "Report Questions the Value of Color Coded Warnings."
39 David Johnston and David Stout, "Bin Laden Is Said to Be Organizing for a U.S. Attack," *New York Times,* July 9, 2004.
40 Tabassum Zakaria, "CIA's Acting Chief Says Threat Highest Since 9/11," Reuters, July 13, 2004.
41 "Kelly: New York Is the No. 1 Target," wcbs880.com, July 13, 2004.
42 Author's interview with Monica Gabrielle, July 11, 2004.
43 "Homeland Security Confirms Election Delay Talks," *Voice of America News,* July 12, 2004.
44 Michael Isikoff, "Exclusive: Election Day Worries," *Newsweek,* July 11, 2004.
45 "Kelly: New York Is the No. 1 Target," *Voice of America News.*

46 Eric Schmitt, "Admitting Intelligence Flaws: Bush Stands by Need For War," *New York Times,* July 10, 2004.

47 Douglas Jehl, "U.S. Sees Evidence of Overcharging in Iraq Contract," *New York Times,* December 12, 2003.

48 Jeff Gerth and Don Van Atta Jr., "In Tough Times, a Company Finds Profits In Terror War," *New York Times,* July 13, 2002.

49 Laura Rich, "On the Job in Iraq and in the Glare," *New York Times,* May 23, 2004.

50 Larry Margasak, "Dems: Halliburton Overcharging for Gas," *Newsday,* October 15, 2003.

51 "Coalition Records Its 1000 Death in Iraq," CNN.com, July 9, 2004.

52 Jim Krane, "Iraq Insurgency Larger than Thought," Associated Press, July 9, 2004.

53 "Report Card on the Occupation," *New York Times,* June 29, 2004.

54 Editorial, "Abu Ghraib, Stonewalled," *New York Times* June 30, 2004.

55 Eric Schmitt, "Congress's Inquiry into Abuse of Iraqi Prisoners Bogs Down," *New York Times,* July 16, 2004.

56 Richard Clarke, "Honorable Commission, Toothless Report," *New York Times,* July 25, 2004.

57 Author's interview with Lorie van Auken, July 14, 2004.

AFTERWORD

1 Valerie E. Caproni, official FBI bio, http.//www.fbi.gov/libref/executive/caproni.htm.

2 Greg B. Smith, "Terrorist Called Pals on Feds Line," New York *Daily News,* September 24, 2000.

3 Patrick Fitzgerald, official DOJ bio, http://www.usdoj.gov/usao/iln/patrickjfizgerald .html.

4 On July 14, 2004, the author sent a detailed letter to U.S. Attorney Ftizgerald setting forth a series of question. In reply, on July 15, Randall A. Samborn, an assistant U.S. attorney and public information officer for the Northern District of Illinois, wrote: "I am in receipt of your letter dated July 14, 2004. Mr. Patrick J. Fitzgerald, United States Attorney for the Northern District of Illinois, respectfully declines your request to answer the question you posed. Thank you for the opportunity."

5 Robert E. Kessler, "FBI Siezes Writer's Data," *Newsday,* April 19, 1997.

6 Robert E. Kessler, "Feds Indict 2 in TWA Theft," *Newsday,* January 8, 1998.

7 Robert E. Kessler, "2 Free on Bail in Flight 800 Theft," *Newsday,* December 10, 199[illegible].

8 Robert E. Kessler, "Couple Convicted in TWA Theft," *Newsday,* April 14, 1999.

9 Robert E. Kessler, "Couple Receive Probation in Theft of TWA Evidence," *Newsday,* July 17, 1999.

10 Christine Negroni, *Deadly Departures: Why the Experts Failed to Prevent the Twa Disaster and How It Could Happen Again* (New York: Cliff Street Books/HarperCollins, 2000).

11 Kristina Borjesson, *Into the Buzzsaw* (Amherst, NY: Prometheus Books, 2002).

ACKNOWLEDGMENTS

Much of the evidence cited in this book with respect to Greg Scarpa Jr., the intelligence he collected from Ramzi Yousef, and its relation to the TWA Flight 800 crash came to me in April, just months before this book was due for delivery. The shocking revelations resulted in a major course correction in the manuscript. So my first debt of thanks goes to Cal Morgan, the extraordinary editorial director at ReganBooks, who had guided me through the editing of my last book with HarperCollins, *1000 Years for Revenge.* Once again his tireless work ethic and attention to detail has resulted in a book that lays out a complicated story in a way that I trust is not only accessible to readers, but backed up with nearly eight hundred end notes and citations.

Another profound expression of gratitude goes to Cassie Jones, the managing editor of ReganBooks. For *1000 Years for Revenge,* she was instrumental in supervising the design and execution of a 32-page illustrated time line in the middle of the book that would have been a challenge for the editor of a news magazine, let along a publishing company, but she did it under incredible time pressure. And with *Cover Up* she has done it again, coping with the detailed illustrated appendices in this book with the same aplomb with which she helped shepherd the text—but on an even tighter deadline.

As mentioned throughout, I was blessed in this stage of my research with access to Paul Thompson's remarkable time lines from the Center for Cooperative Research. Each citation in that database is supported by a news story from the mainstream media. Nonetheless, to be used as source backup, the material still needed vetting. To help me with that daunting task, I called upon an old friend. Years ago, Mary Lou Pizzarello and I had worked together at ABC News. She and her husband Roy, a cardiologist, had generously thrown a party for me after a signing for my last book at Barnes & Noble's flagship store on Fifth Avenue in Manhattan.

Afterward, Mary Lou had asked me if I needed any help with the research on this new book. When I took her up on the offer I'm not sure she understood the full dimensions of the assignment, but I'm forever grateful for her offer, and the work she did in following through on it. On several occasions during my research the Cooperative Research site had gone down, and I was fearful that Thompson's research might be lost. So Mary Lou agreed to spend the next six weeks tirelessly locating the time line stories, downloading them, copying them, and organizing them. The result was an archive of more than twenty thousand pages. All of this was done at her own expense, in the interest of preserving these historic news pieces, which lay out the record of the 9/11 disaster as it happened.

Any researcher, reporter or scholar with an interest in the war on terror would consider the Cooperative Research timelines a bonanza of open source information. And now the material is available not only online but in book form. The trade paperback edition of Paul's work, *The Terror Timeline,* came about as the result of a meeting I arranged in early June between Thompson and my publisher, Judith Regan. With more *New York Times* bestsellers to her credit than any other single publisher, Judith quickly saw the value in his vast database and agreed to bring Paul's timelines to the public in printed form.

For inspiration during the nine months it took to deliver the first manuscript, I was sustained by the input and determination of the Jersey Girls: Kristen Breitweiser, Mindy Kleinberg, Patty Casazza, and particularly Lorie van Auken, who sent me almost daily e-mails with fresh articles to support my research. Another tremendous resource was Monica Gabrielle, who runs the Skyscraper Safety Fund and who, like these other widows, has been transformed since 9/11 from the wife of a businessman into a self-taught intelligence investigator and lobbyist for the cause of full disclosure.

The most important single source for the book was Angela Clemente, a single mother of three who has emerged as one of the top forensic investigators in the country. It was Angela, with her partner Dr. Stephen Dresch, who first became aware of the Scarpa intelligence from Yousef. Digging further, they unearthed the story of Detective Joe Simone, who was set up by the Justice Department to cover the leaks that resulted from the shadowy relationship between Greg's father and Lin DeVecchio.

Before they came to me, Angela and Dr. Dresch had sent a detailed summary of their revelations to the House Government Reform Committee and the 9/11 Commission. When the Commission turned a blind eye to their findings, Angela approached me "over the transom," after reading *1000 Years for Revenge.* "Going to an investigative reporter was a last-ditch effort to get the truth out," she admitted in one of our early conversations. "If Congress and the Commission had done their job, we wouldn't have had to."

During the late fall and early winter, I got further insights into the Commission from a confidential source on the staff who had contacted me in November after reading my book. Several other people who had read *1000 Years for Revenge* contacted me via my Web site, www.peterlance.com, and sent me important material. One of them was Rob Maine, a computer animator from Pasadena, California, who had done some significant open source research on NORAD.

On my last book I received tremendous help from Bob Torragrossa, a retired Army signals intelligence specialist who lost his son-in-law on 9/11, and Joe Murphy, a veteran private investigator from Marblehead, Massachusetts, with whom I had worked years ago on a Chicago arson investigation when I was a correspondent for ABC News. Joe put in hours of pro bono research time looking into a possible al Qaeda sleeper still living in the United States.

This time I got tremendous support in the vetting process from my former attorney and fellow Rhode Islander Jeff Feldman, and from Joseph R. Bailer, a veteran political strategist whom Jeff and I had worked with more than three decades ago while college students at Northeastern University in Boston.

Three other lawyers provided me with the breathing room I needed to get the book done: Professor Jay Carlisle of the Pace University School of Law; John Moncrief, a Manhattan attorney; and Gary Olsen, on the West

Coast. I also owe a debt to the legendary journalist A. J. Weberman, who has spent the past several years exploring the vast body of al Qaeda-related cases that emerged from the PBI's New York office.

As before, this investigation benefited from the tremendous support of my family and friends, especially my son, Christopher, my two daughters, Mallory and Alison, and my father, Joe, who turned ninety when the manuscript was half finished. An ex-navy chief who served on the U.S.S. *Arkansas* during the Normandy invasion, he has inspired me to tell the truth since my first job as a fifty-dollar-a-week cub reporter for the *Daily News* in my home town of Newport, Rhode Island. My cousin Sheila Tyler, a retired nurse, and her husband Harl, along with a neighbor, Anna Friend, have spent years helping my father during periods when my work took me far away from Newport. I can't thank them enough.

Meanwhile, this investigation required that I spend most of the first half of 2004 in New York City, and the man who watched over my office in California was Win Collins, a trusted friend and decorated army veteran. Win had read and offered notes on my last book along with Lucy Kohansamad, another Santa Barbaran, who volunteered to proofread the manuscript at a time when I was really under the gun.

Finally, a special note of thanks goes to Suzanne Merrill, a devoted friend who has lived in London the past few years. She believed in the importance of this investigation and offered tremendous support during the research and writing process. Without her generous help, this book would have never seen the light of day.

PETER LANCE

New York City
August 2004

INDEX